ARTIFICIAL INTELLIGENCE
AND
SIMULATION

ARTIFICIAL INTELLIGENCE

AND

SIMULATION

Edited by

Willard M. Holmes

THE SOCIETY FOR COMPUTER SIMULATION

(Simulation Councils, Inc.)
San Diego, California

Rosemary A. Whiteside
Managing Editor/Art Director

CONTENTS

	Page	Authors

Introduction

'The obstacles to discovery: The illusion of knowledge, the battle against the current facts, and the dogmas of the learned," Daniel J. Bourstine, *The Discovers,* Random House 1983.

The first Artificial Intelligence conference to be held at the 1985 Eastern Simulation Conferences is a new opportunity for expanding the use of simulation in the service of society. I have been involved with basic research in computer modeling and simulation for two decades or more and I have discovered that the thread of continuity that runs through these disciplines and makes them successful is learning. This includes learning from mistakes, from false starts, from my peers, and on rare occasions, from successful research efforts.

Learning and gaining knowledge is a hallmark of all research. Applied research focuses on solving problems related to a particular application whereas the most successful problem solving efforts involve having both the opportunity and the willingness to examine unconventional and non-traditional solution possibilities. The synergism of ideas, concepts, and technologies from different scientific disciplines provides unique opportunities for a non-traditional solution to traditional problems in any scientific field.

In the field of computer science there has emerged a relatively new area of interest called Artificial Intelligence (AI) which has demonstrated a high potential for applications to problems in both the technical and social arenas. The term "Artificial Intelligence" was first introduced by John McCarthy to the academic community. Significant effort was devoted to the search for broad and general laws that would encompass the major concerns of the new science of AI and for programs that could play the perfect chess game.

In the 70's, a dramatic change took place in the direction of research and development of AI. Robotics, vision technology, pattern recognition, natural languages, knowledge engineering, and expert systems are but a few of the disciplines resulting from AI research. Application of AI technologies has contributed to the development of expert systems in the following specialized areas: medicine, mathematics, chemistry, geology, and computer engineering. These are examples of proven and productive fields using expert systems.

Some advocates support the view that AI will be a tidal wave contributing to problem solutions in many areas of society. Others take the position that the influence of AI technology will be more like a rising tide with considerable debate on 'How high the tide?' The most successful application of AI technology to date is in the area of expert systems. In the areas where AI methods have been introduced, the result has been to re-think the approach to operations, to problem solving, and to re-evaluate the tools and techniques used in research. This involves learning from both the specialized domain and AI technology perspective in application to the real world problems.

This became the practical approach in AI after the early 70's when Feigenbaum, after laboring for years to produce the expert system, Dendral, threw a challenge to the AI community in a lecture at Carnegie-Mellon. "You people are working on toy problems," he said. "Chess and logic are toy problems. If you solve them, you'll have solved a toy problem. And that's all you'll have done. Get out into the real world and solve real world problems."

How does this impact or affect the art and science of simulation? It would be difficult to identify areas in our technological society that simulation has not touched or not been of service. Simulation is one of the most powerful analytical tools available to those responsible for the design, development, and operation of complex processes or systems. While computer modeling is heavily based on mathematics, probability, and statistics, the modeling and experimentation remains very much an intuitive process. John McLeod gave another perspective by saying, "It has been said that writing imposes a discipline on thinking; modeling takes the process one step further by quantifying our thinking and establishing the interaction of related thoughts; thus, an important use of simulation is teaching" (or learning).

The interest in Artificial Intelligence by the simulation community is indicated by an increase in both the number of conferences on AI topics and by the participation in AI-related Professional Development seminars sponsored by The Society for Computer Simulation. One objective of this conference is to bring together practitioners, theoreticians, and potential users of Artificial Intelligence and simulation and to focus their attention on existing accomplishments and technology that are applicable to diverse areas of concern.

The papers scheduled for the technical sessions and which are included in this Proceedings feature topics that show how simulation is used to enhance the AI system and inversely, how AI is used to achieve objectives through simulation. Panel discussions covering various AI topics are also included to present a source of new ideas, new names, and new challenges.

Willard M. Holmes, PhD
Conference Chairman

An expert system for chemical process control

A.S. ELMAGHRABY, V. JAGANNATHAN, AND P. RALSTON
ENGINEERING MATHEMATICS AND COMPUTER SCIENCE
UNIVERSITY OF LOUISVILLE
LOUISVILLE, KY 40292

ABSTRACT

A common problem for various engineering disciplines is the study and control of dynamic systems. The tools developed for dynamic systems have proved useful in a variety of other disciplines. This study integrates concepts drawn from the areas of simulation, artificial intelligence and chemical process control to develop an intelligent control strategy for a dynamic system.

The actual process considered is a distillation column. This well known process is selected for two reasons: first, extensive literature concerning control strategies and dynamics is available, and second, the distillation process consumes 30% of the energy consumption in the chemical industry. The practical need to minimize energy usage makes this process ideal for innovative work concerning proper control strategies. The dynamic and steady state simulations for such a process are based on material and energy balances over the column and suitable equilibrium and hydraulic relationships.

The knowledge-base module essentially includes representation of available knowledge needed for the control process. It includes two types of knowledge, one which is independent of the specific column considered and the other dependent on it. Thus, this study investigates the use of an expert system to control a chemical process, by simulating the process. Some preliminary results are included and a methodology is presented including the potential of a learning expert system.

INTRODUCTION

Development of an expert system for any application is only limited by the availability of expert knowledge. A novel approach for inferring such knowledge using simulation techniques is used in this work. An exploratory chemical process application to demonstrate this approach is distillation control.

A common problem for various engineering disciplines is the study and control of dynamic systems. The tools developed for dynamic systems have proved useful in other disciplines.

Presently, process control can be categorized either as classical control or modern control. Classical control refers to the heuristic combining of various levels of single-input/single-output controls based upon deterministic or empirical process models. Modern control refers to the combining of control inputs and outputs in order to provide improved dynamic performance or time optimal responses. Techniques used in modern control are also applicable to non-linear systems with time-varying parameters. Work is proceeding in adaptive control where strategies are developed which can be modified on-line to account for changes in process parameters [1]. The expert system in this application will control a binary distillation column.

Computer modeling and simulation has been used to study and analyze the dynamics of complex systems. Furthermore, when actual experimentation is undesirable, simulation models have been used in lieu of the actual systems. Chemical process control has benefited from the use of simulation models for experimentation with various control schemes.

Expert systems attempt to mimic human ways of reasoning about problems using an explicit body of knowledge in a specific domain. Use of expert systems has grown considerably and is recognized to have a potential for a wide variety of applications in military and industrial environments.

This study explored the feasibility of utilizing simulation models for inference and representation of knowledge required by an expert system for process control. Integration of concepts drawn from the areas of simulation, artificial intelligence and chemical process control is required.

METHODOLOGY

An overall view of the system under study is depicted in Figure 1. The system components are discussed later.

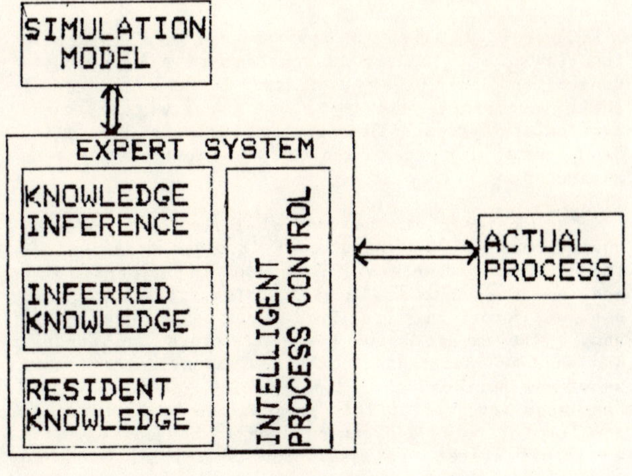

Figure 1. System Components

Expert systems have been used in a diverse set of applications [2,3]. Although strategies for building expert systems are becoming standard [4], there are as many variations in implementation as there are researchers in this field. Typical expert systems are built to assist in decisions in diagnosis or in selecting a particular course of action. Usability is enhanced by generally allowing features such as explanation and natural language interfaces.

Traditionally, controllers have been hardware devices. With the availability of inexpensive microprocessors, the more flexible software implementation is becoming attractive. Use of software in implementing control algorithms can be enhanced by using expert systems. Each application sets some unique requirements on an expert system and process control is no exception.

The following properties are pursued in this research as desirable features in an expert system for control of a chemical process:

1. Providing for appropriate control action to maintain stable operation.

2. Provide for appropriate control action to maintain economic operation in some optimal sense.

3. Diagnose abnormal conditions and provide advisory actions in addition to its control features.

Possible control philosophies using an expert system have been investigated in this study. Feasible scenarios include:

1. Off-line use of the expert system as an advisory system for diagnosis of unusual situations.

2. On-line use of the expert system can be implemented in two distinct ways:

    a. Provision of all the desirable control actions as well as providing for warnings and diagnostics.

    b. Provision of warnings and diagnostics only, with the control actions delegated to traditional controllers.

Control actions for any process are typically functions of inputs, disturbances and outputs. Controllers are highly affected by the choice of their parameters, and these are usually tuned to a particular system. Sensitivity analysis has often been used for the tuning of control systems parameters.

The designer of the control system (the expert) goes through a learning process in order to fine tune the control parameters. This learning process will add to the general knowledge (expertise) of the designer providing specifics about the problem at hand. The expert knows what parameters are tunable and what are constants to a given situation; and this knowledge guides the fine tuning process. The knowledge involved in this process can be captured in the form of rules and can be used in conjunction with an actual system (or simulation of it) to learn to control the given system.

The knowledge-base module (Figure 1) includes representation of available knowledge needed for the control process. Two types of knowledge are considered relevant for this application:

1. Process independent (general) knowledge.

2. Process dependent (specific) knowledge.

Process independent knowledge is a representation of the physical laws governing distillation control. This type of knowledge is also referred to as resident knowledge or seed knowledge. Process dependent knowledge is specific to the particular chemical process under control and is not required a priori by the expert system.

Knowledge inference, the long term goal of this study, is based on learning by controlled experimentation using a simulation model. The inference module will utilize the resident knowledge to control the simulation experiments. The results of such experiments will be screened by the inference module and used to infer system specific knowledge. Representation schemes of this knowledge could utilize several schemes such as production rules, frames or functional relations.

The intelligent (expert) process controller uses the available knowledge, resident and inferred, to control the actual process. It is essentially an expert system based on action inference from the available body of knowledge.

CURRENT IMPLEMENTATION

The actual process considered is a binary distillation column. This well known process is selected for two reasons:

1. Extensive literature concerning control strategies and dynamics is available.

2. The distillation process consumes 30% of the energy consumption in the chemical industry [5].

The practical need to minimize energy usage makes this process ideal for innovative work concerning proper control strategies.

The distillation column has five manipulated (input) variables and five controlled (output) variables. The two disturbance variables are feed rate and feed composition. The manipulated variables are cooling medium flow, reflux flow, distillate flow, bottoms flow, and heat input to the reboiler. The controlled objectives are to maintain one or both compositions at specified value, which requires the pressure to remain constant. For stable operation, suitable levels must be maintained in the base and accumulator. For this application, a conventional dual composition control scheme will be used. Pressure is assumed to be held constant by some manipulation of heat duty and does not enter further in the simulation or expert system. Reflux accumulator level is controlled by distillate flow, and column base level is controlled by bottoms flow. The top and bottom compositions are controlled by manipulation of reflux and vapor boilup flows respectively (Figure 2).

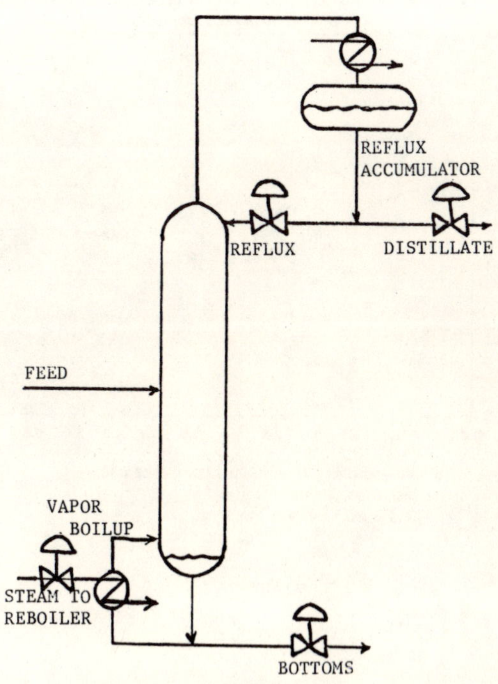

Figure 2. Typical Distillation Column

The dynamic simulation for binary columns used in this work is discussed by Luyben [6]. The system simulated is the binary distillation column example also in Luyben (1973), where the interested reader may find all details concerning the assumptions and simulation development. The dynamic simulation is based on a total and component material balance per tray, a liquid hydraulic equation, and a vapor liquid equilibrium equation. The simulation logic steps are:

1. Calculate vapor compositions on all trays.

2. Calculate liquid flow rates on all trays.

3. Evaluate derivatives.

4. Integrate with Euler technique.

The program is also equipped with proportional-integral feedback controllers to give closed loop response.

Following preliminary analysis, the knowledge needed for control of the ample system was discussed and documented in the form of "If Then" rules. A simplified set of rules is presented in Appendix A.

Initially the rules were used in an off-line mode in conjunction with a Lisp-based inference engine. This general tool has been developed and presented elsewhere [7,8]. Using this preliminary approach, the rules were refined. The need of interfacing to the actual process, or preferably the simulation, made a FORTRAN implementation desirable.

A FORTRAN simulation of the process was adapted. The knowledge involved in controlling the process was coded in the form of rules. These rules are equivalent to those depicted in Appendix A. A simple forward chaining inference scheme was coded for the given rules in FORTRAN. The rules, now representing an expert controller, interacted with the process simulation providing the needed control actions.

An implementation decision was to control the continuous process in a discrete fashion. This has been decided to provide results corresponding to a microprocessor-based on-line control.

The rules used in implementing the controller are general in nature. These rules are pertinent to any column that can be controlled using a conventional control strategy. Values of thresholds, margins of measurement and levels of control action has been determined by a quick trial and error approach, but plans for a knowledge inference scheme that will provide the system with learning capabilities are in progress.

The results of testing the expert system with minimum tuning were encouraging. Comparisons to open loop response and to a traditional (perfected) PI-controller demonstrates the system's ability to control top and bottom compositions, Figures 3 and 4. Based on absolute error criteria, the expert controller is not performing as good as the well-tuned PI-controller; but with the implementation of learning capabilities, this will be improved.

The encouraging feeling about these results stems from a comparison of the control actions of both controllers, Figure 5 and 6. In these figures, the anticipatory nature of the expert controller is clear particularly compared to the reactionary control actions of the PI-controller. The expert controller raises and lowers the values of boilup rate. This is done in anticipation based on its knowledge of the projected effects of an increase (step response) in feed composition. Effects in the manipulation of vapor rate are less outstanding but can be demonstrated for other types of disturbances.

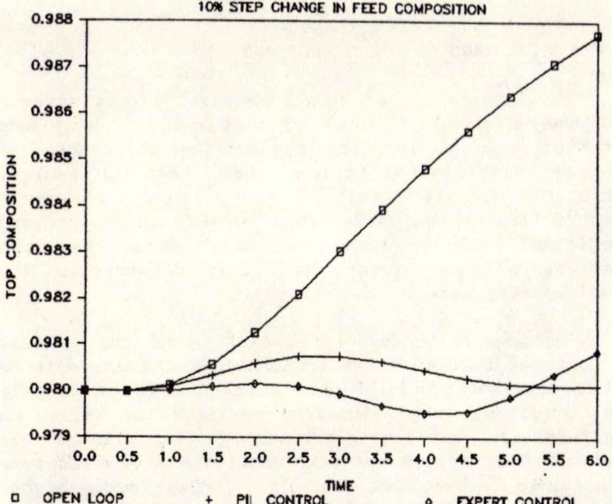

Figure 3.

3

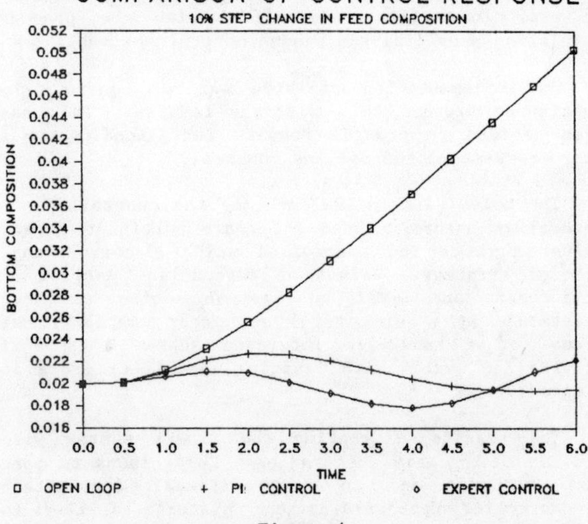

Figure 4

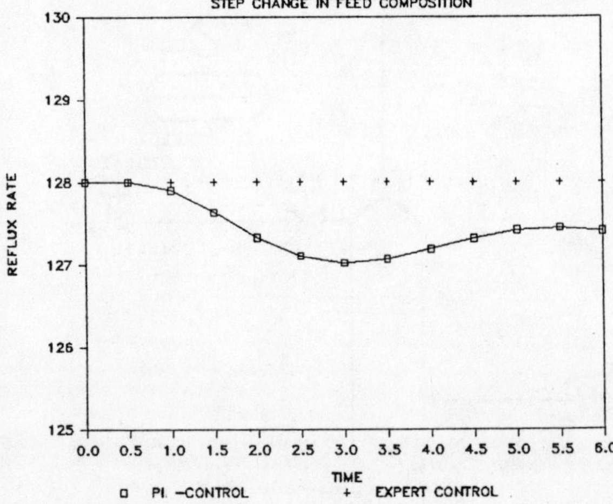

Figure 6

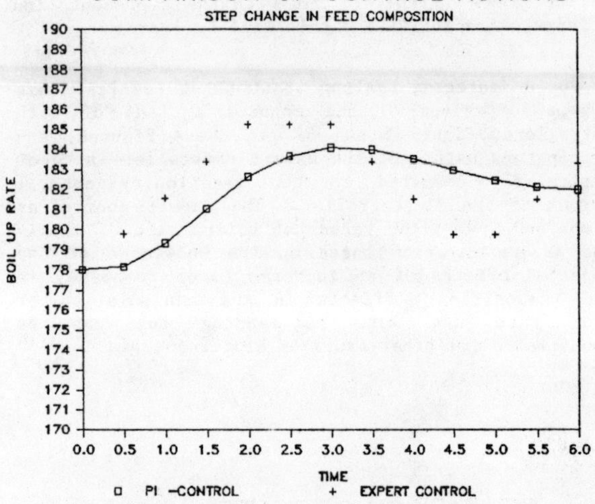

Figure 5

CONCLUSIONS AND FUTURE DIRECTIONS

The application of knowledge-based expert systems to the area of chemical process control has been studied. An exploratory application to control a binary distillation column has been presented. Results show the ability of a discrete expert controller to maintain the levels of controlled variables for a step response. More important actions of the expert controller demonstrate its anticipatory nature.

The use of a dynamic simulation of the process interfaced to the controller provides the convenience of experimentation. Future directions rely heavily on utilizing the simulated process to allow a self-learning scheme. Learning is viewed as inference of system specific knowledge needed to tune the general body of knowledge. Also, some higher level of rules about rules, meta-rules, are being considered to allow the expert system to possibly use the currently implemented conventional control strategy or select other strategies such as material balance control.

REFERENCES

[1] Edgar, T. F. and Schwanke, C. O., "A Review of the Application of Modern Control Theory to Distillation Columns", *1971 Joint Automatic Control Conference*, Vol. 2, p. 1370.

[2] Barr, A., Feigenbaum, E. A., editors, "The Handbook of Artificial Intelligence", Volume 1, William Kaufmann Inc., Los Altos, California, 1981.

[3] Barr, A., Feigenbaum, E. A., editors, "The Handbook of Artificial Intelligence", Volume 2, William Kaufmann Inc., Los Altos, California, 1982.

[4] Hayes-Roth, F., Waterman, D. A., Lenat, D. B., editors, "Building Expert Systems", Addison-Wesley Publishing Company Inc., 1983.

[5] Ryskamp, C. J., "Explicit Versus Implicit Decoupling in Distillation Control", *Second Engineering Foundation Conference on Chemical Process Control*, Sea Island, Georgia, January 1981.

[6] Luyben, W. L., *Process Modeling, Simulation, and Control for Chemical Engineers*, McGraw-Hill, New York (1973).

[7] Jagannathan, V., Elmaghraby, A. S., "Computer-Aided Learning Tool for Simulation", Proceedings of the Third Annual Workshop on Interactive Computing: CAE: Electrical Engineering Education, October, 1984.

[8] Elmaghraby, A. S., and Jagannathan, V., "An Expert System for Simulationists", Proceedings of SCS Multiconference 1985.

APPENDIX A. - SIMPLIFIED SET OF RULES

```
 1.  If current value of feed composition
         is larger than previous value by a margin
     Then, feed composition has increased.
 2.  If current value of feed composition
         is smaller than previous value by a margin
     Then, feed composition has decreased.
 3.  If current value of feed rate
         is larger than previous value by a margin
     Then, feed rate has increased.
 4.  If current value of feed rate
         is smaller than previous value by a margin
     Then, feed rate has decreased.
 5.  If current value of reflux rate
         is larger than previous value by a margin
     Then, reflux rate has increased.
 6.  If current value of reflux rate
         is smaller than previous value by a margin
     Then, reflux rate has decreased.
 7.  If current value of boilup rate
         is larger than previous value by a margin
     Then, boilup rate has increased.
 8.  If current value of boilup rate
         is smaller than previous value by a margin
     Then, boilup rate has decreased.
 9.  If feed composition has decreased
         or, feed rate has decreased
         or, reflux rate has decreased
         or, boilup rate has increased
     Then,  top and bottom compositions are projected to decrease.
10.  If feed composition has increased
         or, feed rate has increased
         or, reflux rate has increased
         or, boilup rate has decreased
     Then,  top and bottom compositions are projected to increase.
11.  If top composition will decrease
         and, its value is at a threshold
     Then, use control action 'a'.
12.  If top composition is low, and is below a threshold
     Then, use control action 'a'.
13.  If top composition will increase, and is at a threshold
     Then, use control action 'b'.
14.  If top composition is high, and is above a threshold
     Then, use control action 'b'.
15.  If bottom composition will increase, and is at a threshold
     Then, use control action 'c'.
16.  If bottom composition is high, and is above a threshold
     Then, use control action 'c'.
17.  If bottom composition will decrease, and is at a threshold
     Then, use control action 'd'.
18.  If bottom composition is low, and is below a threshold
     Then, use control action 'd'.
19.  If action is 'a'
     Then, increment reflux
20.  If acton is 'b'
     Then, decrement reflux rate.
21.  If action is 'c'.
     Then, increment boilup rate.
22.  If action is 'd'
     Then, decrement reflux rate.
23.  If action is not 'a', and not 'b', and not 'c', and not 'd'
     Then, no action is needed.
```

A demonstration of an ocean surveillance information fusion expert system

Elizabeth H. Groundwater
Science Applications, Int., Corp.
1710 Goodridge Drive T-10-4
McLean, VA 22102

ABSTRACT

The author and her team have developed a demonstration ocean surveillance information fusion expert system under an Internal Research and Development (IR&D) program at Science Applications, Int.,Corp.(SAIC). Specifically, the expert system models the thought processes of an ocean surveillance watch analyst attempting to assess vessels' missions and destinations given their correlated tracks, history, and location/status of other vessels in the domain of interest. For the demonstration, rules were developed for determining the mission and destination of selected survey ships.

The expert system was developed using the OPS5 production system and the Franz Lisp programming language on a VAX/VMS computer system. By the end of 1983, a simple demonstration was running. It was then ported to a Symbolics 3670 Lisp machine and the Zetalisp language. The scope of the system is currently being expanded to handle multiple vessel tracks and weather information. The architecture and design of the expert system will be presented in this paper, implementation issues will be discussed, and progress and future plans will be overviewed.

DESCRIPTION OF PROBLEM

Science Applications, Int., Corp. (SAIC) is involved in a number of system development and scientific research contracts to develop sophisticated automated analytical aids and simulation models for users in various disciplines. A few years ago, we began investigating AI/Expert System technology as a means for building these analytical aids and models in areas where human experts are in great demand and short supply. The expert system described in this paper is one of many SAI efforts to develop automated analytical aids which provide expert assistance to analysts, in this case, ocean surveillance analysts.

The U.S. Navy has been engaged for a number of years in the important, voluminous task of surveying and assessing the daily activities of platforms on, under and above the world's ocean surfaces. Information fusion, the merging of different types of information on the same subject from different sources, is a major portion of the problem solving paradigm in ocean surveillance. Information from a multitude of sources flows into ocean surveillance collection centers in huge amounts. There are varying degrees of completeness, reliability, and usefulness of the information, and different reports on the same subject may collaborate or conflict with each other. Many times, there is insufficient manpower to organize, correlate and prioritize the information and too few expert ocean surveillance analysts to interpret and assess the information in a timely manner. This is the problem area the expert system described in this paper has been designed to address.

There are three main stages in the analysis of information about ocean platforms and their movements (tracks). The first stage, after information is collected from sensors, is the determination of whether a sensor has actually seen a platform and the translation of sensor outputs into platform identifications and position reports, if possible. Applications of artificial intelligence techniques in image processing to this stage are currently being researched. Some examples are systems which attempt to pattern-match sensor images of platforms against a stored library of known images to attempt to identify the platforms. Advanced Information & Decision Systems, Inc. has been involved in automatic ship classification from radar signals[1]. SAI's Tucson, AZ office is working on an expert system for identifying target clusters from SAR imagery[2].

The second stage is the correlation of position reports and associated platform characteristics into coherent tracks of platform movements. Problems arise when position reports are incomplete (such as the platform is not identified), unreliable (being off by possibly tens of nautical miles), or ambiguous (when tracks intersect). A number of groups are developing probabilistic or expert system tracker/correlators to solve this stage of the problem. These include SAI with its Bayesian probability-based ISATS tracker/ correlator, Naval Ocean Systems Command (NOSC) with its STAMMER tracker/correlator, - an expert system[3], and the U.S. Navy Center for Applied Research in Artificial Intelligence (NCARAI) with its unnamed correlation expert system currently under development[4].

The third stage involves the analysis of the correlated tracks in the ocean scene to determine destinations, arrival times, and missions of platforms in order to alert U.S./Allied forces, when needed, and keep commanders apprised of the situation. Not much AI/expert system work is being done in this stage, because efforts have focused more on the

previous two stages. However, this stage
suffers from the same problem as the others:
too much data and too little experienced staff
to deal with the data. So, the need for intelligent automation exists here too. What work
has been done is primarily in the areas of
intelligent data management and display in Verac
Inc.'s INCA system[5], allowing human analysts to
spend more time analyzing the data and less time
getting it in a form agreeable to analysis.

We chose to concentrate our development of a
demonstration expert system on this third stage
because of the standing requirement for this
type of information and the lack of previous
research. The demonstration chosen was to
simulate how intelligence analysts solve the
problem of determining the probable destinations
and missions of selected survey ships. A number
of factors led to this choice:

o   The ability to develop an unclassified
     system to solve this problem;

o   The relatively small, workable number of
     survey ships, stations and missions,
     allowing a meaningful demonstration
     system to be developed in less than a
     year; and

o   The availability of an in-house expert
     ocean surveillance analyst familiar with
     survey ship operations to provide the
     needed expertise.

The goal was to show feasibility of the expert
system approach to solving one aspect of the
ocean surveillance information fusion problem
within a short timeframe and demonstrate that
the approach can be naturally extended to other
aspects.

<u>EXPERT SYSTEM ARCHITECTURE</u>

The expert system is an event-driven system,
given platform tracks and updates on platform
tracks, it makes new assessments or updates past
assessments of track destinations and
missions. It is not a goal-driven system, such
as one which has a goal of making a diagnosis,
then searches for information supporting the
diagnosis. The system must be able to operate
in real-time and deal with time issues in order
to have any practicality, once implemented.
Thus, it was designed to read in event
information from a scenario file (for
demonstrations) or a tracker/correlator system
periodically, make assessments, display its
results, then cycle back to receive more information. Its operation, therefore, is very
different from the typical expert system which
only derives one assessment from one unchanging
set of inputs.

The expert system has a production-rule
forward-chaining-based architecture, building
its assessments forward from sub-assessments
built from input events by application of production rules. Production rules are "fired" or
executed only when all the conditions on their
left-hand side have been met, indicating the
appropriate situation exists for application of
the actions on the right-hand side of the rule.

Ocean surveillance analysts bring a number of
sources of knowledge to bear upon a situation
assessment problem, then weight the likely
situations proposed by the different sources of
knowledge against each other to determine the
final solution. Production rules in the expert
system are grouped into separate knowledge
sources which model this process. The five main
knowledge sources contained in the expert system
are:

o   Deployment history - if the platform ID
     or type is known, estimates likely
     destinations and missions based on past
     deployments of that platform or platform
     type in a ship history data base.

o   Survey ship relief posture - for the
     survey ships which are on stations
     currently, calculates the relief posture
     of those ships based on their arrival
     time at the station, the current date/-
     time, and the nominal amount of time
     spent on that station, then estimates
     station destinations likely to be due
     for relief soon.

o   Other ships' activities - determines if
     U.S./allied activities are present and
     are the type of interest to the survey
     ships, then calculates intersect times
     and distances and assesses whether the
     ship seems to be moving toward or away
     from the closest point of approach to
     each U.S./allied activity.

o   Ship's own movement - from the platform
     track's current position and course,
     determines the proximity of its position
     to standard historical tracks to survey
     ship stations and estimates likely
     track(s) it seems to be following.

o   Multi-mission resolution - takes the
     likely destinations and missions
     estimated by the previous four knowledge
     sources and derives a combined set of
     estimates by weighting the evidence.

Time is a major factor in the last knowledge
source. As a platform moves from its home port
to its destination, estimates from the knowledge
sources have different relative weights at
different times along that path. For instance,
deployment history has a high weight near the
beginning of the path and ship's own movements
have a low weight, because there is not much
information in a ship's position and course when
it is only a few miles out of port. However,
after a few days, deployment history has a low
weight and ship's own movement has a high weight
because it has shown intent by being located on
or near a historical track to a station or an
intercept track to a U.S./allied activity, and
it does not matter much what it has done in the
past.

<u>DEVELOPMENT APPROACH</u>

At the beginning of the project, the ocean
surveillance expert and knowledge engineer had
many discussions to determine a suitable problem
domain. Also, information was collected on

current ocean surveillance systems and their capabilities and on-going research activities in the area. Once the survey ship problem domain was selected, the expert developed three specific scenarios listing successive observation reports and assessment decisions made by the ocean surveillance analyst. The first scenario was a normative one where a survey ship followed a historical track to relieve a ship on station which was due for relief. The second was non-normative: a survey ship deliberately followed one historical track for a while, then switched to its intended track. The third was a normative one, where a survey ship intercepted a group of U.S. ships to monitor its activities.

The history data base and current situation data base were designed next. Here, NOSC's work in defining the elements of naval domain knowledge[6] provided to be very useful. The history data base contains simulated deployment histories for a few survey ships and platform characteristics, such as maximum and average speed, used by the expert system knowledge sources. The current situation data base contains platform track information, past and current assessments, status of other survey ships on station, status of U.S./allied activity movements, and the current date/time and other status indicators.

The expert and knowledge engineer then stepped through each scenario to expose the heuristic rules-of-thumb used by the expert to make assessments. This was an iterative process requiring steps to be gone over again and again to reveal the details involved in the expert's decision-making process. The production rules which implemented the expert's heuristic rules were then coded and tested against the three scenarios. Liklihood rating strategies were successively devised and modified until the expert system was able to make roughly the same assessment at each step of each scenario as the expert.

The expert system was written in the OPS-5 production-rule language[7] and in Franz LISP[8] (primarily for geometric calculation routines) on a VAX/VMS system. We developed a simple user interface and explanation capability to allow some analyst interaction. The system ran slowly because the Franz LISP code was interpreted vs compiled and the VAX was heavily loaded, so it was ported to a Symbolics 3670 computer.

The initial demonstration expert system was implemented in eight months with an effort of eight man months. It contains 180 OPS-5 rules and 55 LISP routines for a total of approximately 3000 lines of OPS-5 and LISP code. Of the OPS-5 rules, 70 represent true expert knowledge and 110 are "overhead", implementing the real-time iterative assessment process, managing the history and current situation data bases, and providing the user interface display and explanation capabilities.

FUTURE PLANS AND IMPLEMENTATION ISSUES

We are continuing the development of the expert system this year. Currently, the system can estimate destinations and missions for one ship track at a time. We are adding multi-ship mission resolution so the system can handle more than one ship track at a time. Thus, if there are four likely destinations/missions and three ship tracks in a scenario, the expert system will allocate some subset of the likely destinations/missions to each of the tracks. We are also adding the handling of weather information and considering weather avoidance to the decision-making process.

The goal of this year's effort is to demonstrate the full range of factors affecting determination of the destination and mission of one type of platform. From this feasibility demonstration, informed estimates can be made of the effort required to implement a comprehensive system to handle many more types of platforms. Issues important in the implementation of such a system include:

o   Portability - Once the expert system has experienced all its design changes and testing in its LISP/OPS-5 form, it may have to be recoded in a military standard language such as Ada[TM] and/or moved to militarized hardware to be useful.

o   Interconnectivity - The expert system will have to be able to interface with current and planned military tracker/correlator and tactical display systems for its input and output functions.

o   Distributed Processing - The knowledge sources are currently executed serially, but there is no reason they cannot be executed in parallel. A distributed implementation may be considered in cases such as aircraft assessment where speed is essential.

ACKNOWLEDGEMENTS

The expert system would not have been successful, nor this paper possible, without the efforts of the whole project team, including George Futch, our ocean surveillance expert, Chris Kostas and Tim Germer, A.I. programmer/analysts, and Bernie Cotton, quality assurance and advice giver. The work performed on this project was supported by SAI's IR&D program and the resulting algorithms are proprietary to SAI.

REFERENCES

1. Drazovich, Robert J., Wishner, Richard P., and Warnow, Tandy J. "An Artificial Intelligence Approach to Radar Target Classification," Presented at Tri-Service Workshop of Missile Ship Targeting, Naval Research Laboratory, Washington, D.C., 10-12 August 1982.

2. Kruger, Richard P., "Knowledge Based System for Training and Exploitation," Science Applications, Inc., Tucson, AZ, videotape, January 1984.

3. McCall, D.C., et al. "Stammer2 Production System for Tactical Situation asssessment," NOSC Technical Document 298, October, 1979.

4. Davis, Laura C. and Aldrich, James R. "Knowledge-Based Approach to Naval Multisensor Information Integration," Proceedings, Trends & Applications, 1983, May 25-26, 1983, National Bureau of Standards, IEEE Computer Society Press Order No. 472.

5. Gross, Leonard S. and Payne, Harold J. "INCA: An Environment for Exploratory Development of Tactical Data Fusion Techniques," Presented at the CIA Artificial Intelligence Conference, December 14-16, 1983.

6. Bechtel, R.J. "Elements of Naval Domain Knowledge," NOSC Technical Report 705, July, 1981.

7. Forgy, Charles L. "OPS5 User's Manual," Department of Computer Science, Carnegie-Mellon University, Pittsburg, PA, July, 1981.

8. Foderaro, John K. and Sklower, Keith L. "The Franz Lisp Manual," Department of Computer Science, University of California at Berkeley, CA, September 1981.

9. Groundwater, Elizabeth H., "Final Report for an Expert System Demonstration in Information Fusion," Science Applications, Inc., McLean, VA, January 1984.

Artificial intelligence from the systems engineer's viewpoint

by

Robert D. Hawkins
E-Systems
Waples Mills Rd.
Fairfax, VA

ABSTRACT

The use of decision-tree logic and relational database search techniques has come to dominate the thinking in Artificial Intelligence (AI) circles. These methods are generally implemented by digital computer techniques and have led to a significant amount of research in the computer science arena. The contributions of the linear systems techniques; such as linear programming, dynamic programming, adaptive control and optimal control have largely been ignored. This paper attempts to show a rationale for a renewed effort in the application of these methods in the advancement of AI objectives.

In particular, the application of continuous system simulation methods to provide "a priori" information as to the state of a subject system, as well as provide a means of manipulating a semantic network representing a knowledge-based expert system is reviewed. A typical expert system provides for a generalized computational formalism with three major components: a database of assertions which represent the facts for a specific problem or which model the environment; rules that read and write on the database; and a rule interpreter or inference engine that selects the rules to be applied.

The role of simulation is extremely important to the success of the inference engine. Proper structural and functional models of the target system will enable the use of simulations to predict the behavior of military platforms, vehicles, electronic gear and other subsystems. Such expert systems should be able to acquire operational information directly from sensors or other systems instead of relying upon human intervention. The mechanism of control of physical (and non-physical) systems without human intervention has been addressed with some degree of success in the past with the application of adaptive and optimal control methods. This paper suggests methods for the application of those techniques.

INTRODUCTION

In the early days of computer development, and in the repeated false starts of many well-meaning academicians, the once impractical and still controversial facet of computer research that holds a broad and startling promise for the Department of Defense (DoD) was given birth. This child of ignominy is Artificial Intelligence (AI).

Today, conditions have changed such that industry and DoD are committing untold millions of dollars to obtain the yet unfulfilled blessings of this immature technology. Currently, what are known as expert systems (or more approximately, knowledge based systems), utilize AI methods to solve problems and to aid decision-making. Included among these previously difficult to manage tasks are the diagnosis of illness, prospecting for mineral deposits, analyzing investments, and configuring computer systems.

The events that contributed to the state of affairs is dual in nature. First, there are now important problems to solve that have proven to be too complex for current conventional computer technology. Secondly, the recent advent of abundant and cheap computing power (and the promise of even more of it), opens the door for this discipline that has for so long been considered a useless science to finally yield some useful technology.

Background

Although modern concepts of AI are attributed to those current "giants" in the field (Minsky, McCarthy, Newell and Simon, Feigenbaum, et al), the true origins of AI are embedded in the history of the conception and development of the computational engine (or computer). In 1812, Charles Babbage conceived of the idea of a calculating machine to produce mathematical tables free of errors. This he called a "Difference Engine", and even before he could complete it, he had further conceived of an even more advanced engine which he called the "Analytical Engine". However, the technology required to support the development of the analytical engine was not present and Babbage foundered in his quest. Nevertheless, it is the first recorded example of a programmable machine for computational purposes.

During his long quest for the Analytical Engine, Babbage became the recipient of the patronage of Lady Ada Augusta Byron (Countess of Lovelace, Baroness of Wentworth), the daughter of Lord Byron (the poet, George Gordon). In describing the Analytical Engine in 1843, she cautiously raised the question, "can machines think"; and in as cautious a manner, she let the question die. But, it should be remembered that Babbage was the first to conceive of a computing machine that could play chess. As for the Lady Byron, she will be forever remembered as the personage who gave her name to the recent DoD Higher Order Language, ADA, as the world's first recognized computer programmer. This question of "machine thinking" will be found to be present in practically every effort in which mankind has sought to control his environment. The burgeoning field of servomechanisms and mechanical feedback systems gave rise to similar questions in the early 20th century. In 1915, Leonardo Torres Y Quevedo, a member of the Royal Academy of Madrid, is reported to have said, "The inventor claims that the limits within which thought is really necessary needs to be better defined, and that an automation can do many things that are popularly classified with thought." This remark was made with respect to the development of electromechanical machines that could play certain chess end games. The idea of automating processes that normally require human thought had been broached and would forever after be associated with the development of that facet of intellectual inquiry that we now call Computer Science.

Babbage and Lady Byron had introduced the concept of the stored program computer as we know it today, but were unable to produce a satisfactory Analytical Engine because the technology was not present. As technology progressed from the Renaissance to the Industrial Revolution, it became more readily available to the designers of calculating machines. The principles of counting machines had been established by Blaise Pascal in 1642, and Babbage had advanced the art to the automatic solution of mathematical equations by the time of his death in 1871. What was still needed was an articulate definition of the mathematical logic of computing machines. This was achieved with the definition of the "infinite state machine" by Alan Turing in his landmark paper of 1936; entitled, "On Computable Numbers with an Application to the Entscheidungs-Problem."

This abstract machine is known under various titles as the Universal Turing Machine, the Turing Machine, or simply the Infinite State Machine. It can be imagined to have an infinitely long tape, divided into squares, each of which either has one of a finite number of symbols on it or is blank. This tape traverses the machine, which can scan only one square at a time, and is capable of moving the tape in either a forward or backwards direction. It can only move the tape one square at a time, and it can erase a symbol and/or print one. Such a machine would be capable of performing a variety of calculations, and Turing was able to establish and prove several theorems about them. He was able to put some of his theories to the test in early 1940, when the British government recruited a team of top mathematics and electronics experts to break the German Enigma machine code by developing a decryption machine. This machine, called Colossus, was a special-purpose device and therefore, did not fulfill Babbage's dream of an Analytical Engine. However, the war years did spawn the first generation of computers that came to be called "electronic brains".

The Fulfillment of Babbage's Dream

By 1939, the limitations of the counting machine (or electromechanical calculator) was being questioned. In that year, Professor Howard Aiken of Harvard conceived of combining existing calculators, such as the International Business Machines (IBM) 601 Multiplier, into a single super calculating machine. However, he decided that any such project should have a virgin start from basic principles. At the time, Aiken was a Lieutenant in the U.S. Naval Reserve. And since the Navy was interested in the project, it detached him to pursue its development. Thomas J. Watson, then Chief Executive of IBM, contributed $500,000.00 in funding and an engineering team was made available to Aiken's project. Five years later, in 1944, the Automatic Sequence Controlled Calculator (called the Mark I) was unveiled to the public. On the eve of the presentation ceremony, Aiken introduced the Mark I to the press without acknowledging the contribution of Watson or the IBM engineering team in its development. This enraged Watson and he ordered his engineers back to the lab and the drawing board with instructions to devise a machine of such stunning performance that it would completely eclipse that of the Mark I. But, before this could be done, the first fully operational electronic computer was developed at the University of Pennsylvania, in 1946.

In 1942, the Moore School of Electrical Engineering at the University of Pennsylvania received a $400,000.00 contract from the U.S. Army to develop an electronic calculating machine. This machine was designated the Electronic Numerical Integrator And Calculator (ENIAC), and was intended to generate much needed firing and bombing tables for the North African Campaign. The principal designers of the machine were professors J. Presper Eckert and John W. Mauchly. In 1945, they were joined by Dr. John Von Neumann in planning a newer, more powerful machine, labelled the Electronic Discrete Variable Automatic Computer (EDVAC). But EDVAC was not completed until 1951, which was long after Eckert and Mauchly had delivered the ENIAC and broken away from the Moore School to found their own company.

In 1947, Eckert and Mauchly formed the Electronic Control Company to design and build the UNIVersal Automatic Computer (UNIVAC). The company name was later changed to the Eckert-Mauchly Computer Corporation and a contract was written with the Northrop Aircraft Company to design and build the BINary Automatic Computer (BINAC). The BINAC was delivered in 1949 and was the first stored program electronic digital computer. However, Eckert and Mauchly were unable to generate sufficient sales to achieve business success, and finally sold their company to the Remington Rand Corporation in 1950. The first of a long line of UNIVAC machines was delivered in 1951, and in 1955 the Remington Rand Corporation merged with the Sperry Corporation to become Sperry-Rand.

By 1948, the IBM engineering team had produced the Selective Sequence Electronic Calculator (SSEC). Under the pressures of the Korean War and the competition of the UNIVAC, IBM produced the model 701. It was the first of the 700-7000 series of IBM mainframes and was 25 times faster than the SSEC. This is the celebrated line of mainframes that include the 704, 705, 709 and 7090, 7094 models which established IBM as the leader in the digital computer industry. These first generation computers and succeeding generations of computers were primarily general-purpose (GP) mainframes which had the broadest appeal to both business and scientific applications. However, enterprising persons with an eye for the unusual were already beginning to mechanize intelligence in terms of games, voice and pattern recognition, and natural language machine translation by the beginning of the 1950's.

There is still some controversy as to who deserves sole credit for the concept of the stored program computer. Babbage can certainly lay claim to the concept with his use of the Jacquard loom format. There are those who insist that John Von Neumann was the guiding light to the success of the Moore School machine, the first recognized computer. And, of course, the claims of Eckert and Mauchly were substantiated by patent rights. But these were recently disallowed by the Supreme Court in favor of John Atanasoff of Iowa State University. Who will ever uniquely claim to be the originator of the stored program computer? And, what matters who it is? The dream of Babbage has surely been fulfilled and all mankind is the richer for it.

The Birth Of Artificial Intelligence

Thanks to the efforts of the aforementioned individuals, the world was finally gifted with a machine that could perform a succession of numerical tasks without human intervention. This is a far cry from reproducing the complete spectrum of activities of the human brain, but it does represent a modest beginning. The very natural quest to improve the quality of the fundamental numerical processes as well as the responsiveness of the machinery led to the development of the academic discipline referred to as Computer Science. The stage was now set for the assault on the unknown.

During the summer of 1956, the Rockfeller Foundation provided $7,500.00 to fund a meeting for a two-month, ten-man study of artificial intelligence at Dartmouth College in Hanover, New Hampshire. Four persons; John McCarthy (of Dartmouth College), Marvin Minsky (of Harvard), Claude Shannon (of Bell Telephone Laboratories) and Nathaniel Rochester (of IBM) formed the core of the discussions which were a forum to talk about the various works that they had been pursuing toward making machines behave intelligently. These men were joined by another six persons who shared their interests; Trenchard More (also of IBM), Arthur Samuel (again of IBM), Oliver Selfridge (of the Massachusetts In-

stitute of Technology (MIT)), Ray Salomonof (also of MIT), Allan Newell (then of Rand/Santa Monica), and Herbert A. Simon (of Carnegie Institute of Technology). Other persons journeyed to Dartmouth for occasional short visits pertinent to their own works.

The Dartmouth conference represented the beginning of AI as a separate and distinct aspect of computer science. McCarthy's purpose in organizing the two-month forum was to bring together all of the then-serious researchers in the field of AI and to open channels of communications amongst them. A most pleasant surprise at the meeting was the presentation of a work that had been done at Carnegie Institute of Technology (now Carnegie-Mellon University of Pittsburgh) by Allen Newell and Herbert A. Simon. They described a theorem-proving computer program (called the Logic Theorist) which was the first program to use a computer as a symbolic processor, not as a number cruncher. It is now recognized as the first example of an AI program. In collaboration with J.C. Shaw (of the Rand Corporation), Newell and Simon had created a computer language capable of modelling some simple human problem-solving capabilities, and they used it to develop the Logic Theorist. This language, called Information Processing Language (IPL), was the first of several that processed concepts rather than numbers -- the first symbolic processor and a major step in the direction of machine implementation of cognitive thought.

The intended purpose of the Dartmouth conference was never fully attained. It was a relatively free-structured workshop that left many of its participants with unfilled aspirations. The attendees brought their own ideas and opinions with them and stubbornly clung to these prejudices and biases, and seemingly no one converted anyone else to their viewpoint. Nevertheless, it served as a staging area for those persons that have come to be regarded as the focal points for all future AI research. The very name, "artificial intelligence", was coined by McCarthy and adopted at his insistence over the strenuous objection of others that weren't as comfortable with the term, "artificial". For many years since, many workers have referred to their efforts as "machine intelligence", and not "artificial intelligence".

Over the years, a consistent source of AI research funding has been the military. Throughout the 60's and the 70's, the Advanced Research Projects Agency (ARPA) of the DoD was the major source of financial support. The agency recently had its title updated to Defense Advance Research Projects Agency (DARPA) and has mainly concentrated its funding efforts on four centers: Carnegie Mellon University in Pittsburgh, Pennsylvania (Newell and Simon); Massachusetts Institute of Technology in Cambridge, Massachusetts (Minsky); Stanford University in Palo Alto, California (McCarthy); and Stanford Research Institute in Menlo Park, California (which is literally infested with former students or

the Dartmouth Conference participants).
DARPA is currently preparing to invest considerable sums of money in a mixed bag of
research and development covering supercomputer architecture, AI, and software. There is
some 50 million dollars in its 1984 budget,
and the agency is asking for 95 million
dollars for 1985. It is estimated by some
that the project could ultimately cost 500 to
700 million dollars.

AI AS AN ENGINEERING METHODOLOGY

Apart from its' somewhat amorphous beginnings, AI is still a question in the minds
of many. The twin goals of AI are considered
by most to be the answers to the questions:
"What is intelligence"?, and "How can
machines be made to exhibit intelligence"?
Engineering approaches to the answers to
these questions have led investigators to
studies of the human brain in the hope of
imitating its processes. As a result of
these studies, it is known that the brain,
unlike today's digital computers, uses continuous variables rather than discrete binary
variables. It is also known that the brain
operates in a highly parallel, rather than a
serial fashion. That is, it performs many
operations at once; rather than one operation
after another in tandem.

However, "fuzzy set" theory is being
applied to computer software design in the
hope of enabling computers to make distinc-
tions on the basis of continuous variables.
This theory is a concept that some hope will
bring computers closer to the mode of reasoning used by humans. A fuzzy set is one in
which membership is not precisely specified;
for example, the set of "big" investments.
How "big" is big?? But, with fuzzy sets, it
may be possible to quantify human concepts,
such as "small", "big", "young", "old",
"high", or "low" into a form recognizable by
a computer. And it may yet be a long time
before this concept is accepted as being
practical, for there is great controversy
wherever the subject goes.

Much of the controversy revolves about
the approach one utilizes to introduce imprecision into the computing process. Some
maintain that any kind of uncertainty or
imprecision is best handled by probability
theory and that there is little that can be
done by fuzzy sets that can't be done better
by a judicious application of statistics and
probability theory. There are also those who
insist that the theory of fuzzy sets is just
that — a theory. Their assertions are that
the theory has not, and will not, lead to any
demonstrably useful purpose. And, of course,
there are those who are intensely committed
to the deeply entrenched tradition of respect
for that which is precise, rigorous, and
quantitatively representative. These persons
view with disdain and suspicion any theory
that attempts to abrogate the precision of
exactitude and to arrive at an accommodation
with the obvious and pervasive imprecision of
the real world.

AI is currently being applied in many
diverse applications with varying degress of
success. Included among these are:

- Natural Language Translation

- Knowledge-Based Expert Systems

- Robotics

- Computer Programming

Natural language translation is a function that has eluded computer scientists for
many years. One of the most ambitious
natural language research projects is now
underway in Palo Alto, California. The project involves computer scientists, logicians,
linguists and philosophers from Stanford
University, Xerox – Palo Alto Research Center, SRI – International and Fairchild.
Knowledge based expert systems is that part
of AI that has been most publicized because
of the impressive performance of some of
these programs in imitating the way a human
expert makes decisions. Robotics is another
facet of AI that has received a good deal of
attention from the commercial communities.
It is largely concerned with the automation
of product design and manufacturing functions. The DOD in particular, is very much
interested in robotics as a means of substituting machinery for manpower; especially
in hazardous environments. Computer programming is a facet of AI that is expected to
contribute significant advances in making a
computer more "user friendly" by facilitating
the man/machine interface. It is becoming
less concerned with scientific calculation
and more with AI applications. Data processing now handles vast quantities of non-
numeric data; such as symbols, speech,
graphics and images.

Let us consider now that the many
approaches to the development of working AI
systems includes the following emphases:

- Logical

- Deterministic

- Statistical

- Probabilistic

- Heuristic.

Each of the above should be considered
in the development of new AI systems, and
each contributes a measure of insight into
intellectual behavior.

Logical Methods

The utilization of the computer as a
logical device for the purpose of defining a
unique solution to a decision factor falls
within this methodology. This is the classical approach to AI and began with McCarthy's
development of LISP, the currently favored
language for the development of knowledge
based (or expert) systems. These systems are

computer programs that mimic human experts by
the use of computer logic. By using methods
and information acquired from human experts,
expert systems can solve problems, make predictions, and offer advice with degrees of
accuracy approaching that of their human
counterparts.

A typical knowledge-based system consists of an inference engine, a knowledge-
base, and a workspace (see Figure 1).   The
inference engine attacks a problem by searching the domain (or search space) of knowledge
contained in the knowledge base.   The workspace is an area of memory that is set aside
for the storage of a description of the
problem constructed by the system from facts
supplied by the user.

An inference engine is essentially a
computer program designed to process symbols
that represent objects, rather than compute
numbers.   Actually, all computers are symbol

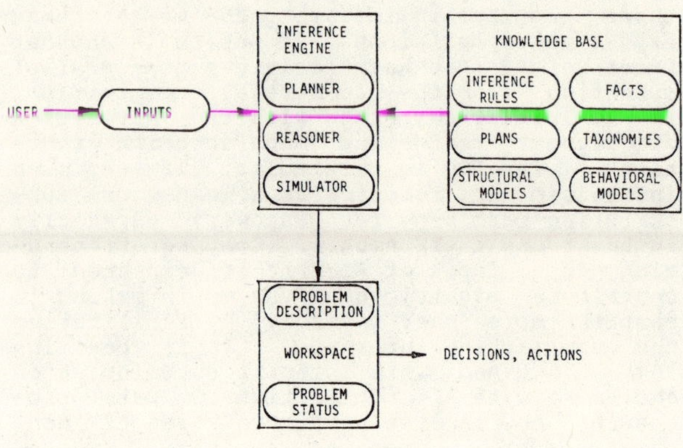

FIGURE 1

A TYPICAL KNOWLEDGE-BASED
EXPERT SYSTEM

processors. However, in traditional computer
applications, symbols primarily represent
numbers and processes are mathematical operations. In expert systems, symbols may represent virtually any type of object; a person,
a concept, a process, a class of objects.
Objects that are considered as elementary by
the   system ("atoms") are represented as
strings of alphanumeric characters ("atomic
symbols"). For example, the character "TANK"
might represent the atomic concept TANK, IS-A
(is a) for the atomic relationship of
membership in a class, and "VEHICLE" for the
atomic class of man-made objects.   Complex
objects   are   represented in memory   by
connected lists of atomic symbols.   For
example, the fact that all tanks are vehicles
could be represented by the concatenated list
TANK IS-A VEHICLE.

The AI computer reasons (or associates)
by processing such symbols. The most important symbol processing functions are: 1)
matching two character strings; 2) joining or

separating two strings; and 3) substituting
one string for another.   Such operations
emulate reasoning.   For example, to find all
assertions relevant to tanks, the computer
would search its memory space for all lists
containing the character string TANK.   A
matching operation, followed by a substitu-
tion operation, and finally a deletion operation would permit a system to synthesize the
fact that TANK IS-A MAN-MADE OBJECT from the
facts TANK IS-A VEHICLE and VEHICLE IS-A MAN-
MADE OBJECT.

Expert systems have been written in many
different computer source languages.   These
source languages are commonly referred to as
procedural languages, primarily because they
define the basic procedures that permit the
user "to address" the computer.   In addition
to these fundamental (procedural) languages
which are sometimes referred to as Higher
Order Languages (HOLs), there are additional
languages that facilitate the user's ability
to develop applications of various sort on
the computer.   These languages are generally
embedded (or written) in the procedural
language chosen for the target computing
machine and are generally referred to as
Problem Oriented Languages (POLs).   They are
primarily used to facilitate the development
of application modules for a distinct technical discipline.   Typical of these HOL/POL
combinations are the languages C and UNIX*.
These languages were developed by members of
the Bell Telephone System for the express
purpose of implementing real-time computer
networks.   However, simulationists may be
more familiar with the FORTRAN/CSSL combination which facilitates the development of
continuous systems simulation modules.

LISP is the most common POL in use in
the AI community.   It was probably originally
written in FORTRAN, which is extremely
limited for use in the development of AI
systems.   An AI programming language must
manipulate symbols that represent more than
just numbers.   The determination of the elemental relationship among symbols is called
association, and in order to associate symbols, a language is needed that differ-
entiates between lists.   LISP (for list processing) was developed in 1958 by John
McCarthy, the organizer of the first AI symposium and the originator of the term "artificial intelligence."

As previously stated, the elemental
symbol in LISP is the atom.   It is similar to
a variable name in traditional procedural
languages.   An atom is named by a character
string and can be assigned a value.   Some
atoms have preassigned values in LISP; for
example, T represents true and NIL means
false.   The value of other atoms can be
assigned by the programmer.   A list is defined as a sequence of elements, each element
being either a list or an atom. This definition is recursive; a technique common to

---

*UNIX *is an "operating system". What is inferred
here is that like most HOLs, it is a pre-processor.
In fact, it is both a pre-processor and a post-
processor.*

compilers, assemblers, interpreters, and other digital computer translators.

Consider the typical list: (THIS (ALPHA, BETA) IS (A) ((LIST)) ()). The first element in the list is the atom THIS; the second element is a list of two atoms ALPHA and BETA; the third is the atom IS; the fourth is a list of one atom, A; the fifth is a list whose only element is a list (containing the single atom, LIST); and the final element, (), is a null list which contains no elements. LISP routines are in themselves lists. This accommodates recursive programming. Recursive techniques are economical in expression. They do not specify an explicit sequence of operations, but define an algorithm stated in terms for the Nth item in terms of its relationship to previously calculated items. For example, consider the factorial of N, which is often defined recursively as:

Factorial of N = 1, if N = 1

Factorial of N=N * Factorial of (N-1), if N>1

This definition translates into LISP as:

FACTORIAL N: (COND ((EQUAL N 1) 1)

(T (TIMES N (FACTORIAL (DIFFERENCE N 1))))).

The function EQUAL takes two expressions (in the above case, N and 1) and returns T if the values are equal. The function COND takes pairs of expressions and returns the value of the right member of the first pair, for which the value of the left member is evaluated as T (true). If we evaluate the above routine for N = 4; FACTORIAL calls itself, but each call is for a lesser number (3, then 2, then 1) until the final call occurs, which returns the value of 1. Applying this procedure, the factorial of 4 is expressed as 4 * 3 * 2 * 1 = 24.

Recursion will be found to be important in expert systems development because many AI tasks exhibit recursive structure. In essence, the same solution process that works with a complex problem can be used on a simple version of the problem, or on a subset of it. In addition, the solution process can be defined to a great extent without knowing the elements of the problem involved. Notice that the LISP routine above is a list. Both LISP programs and data are structured as lists. Any structure, such as an array, for example, can be represented as a list. Although this may not be efficient (in terms of execution time), AI programs must be able to manipulate data structures of unpredictable size and extent that evolve as an expert system runs. A list structure is flexible enough to allow easy manipulation by a programmer as well as the expert system itself. In the latter case, the list data structure makes it possible for one LISP program to process another LISP program as data. LISP programs readily analyze and generate LISP programs; in particular, they modify themselves, which is essential if an expert system is to learn on its own.

A human expert's skill depends largely on the rules he (or she) derives from experience and existing knowledge. And like the human expert it imitates, the expert system is limited by the amount of data it must handle. The larger the domain of interest encompassed by an expert system, the more facts and rules it must assimilate. As the search space increases, the time required to search the space increases. If the domain of interest becomes too large, the efficiency of the system declines to an unacceptable level. Expert systems function best in narrowly defined circumstances in which the alternatives are limited in scope and the search space is limited.

The logic of expert systems is based upon a philosophy of absolutes. This is a form of logic proposed by Aristotle (384 BC-322 BC) and is, therefore, referred to as "Aristotelian" logic. It admits only two states; right or wrong, yes or no, black or white. Such a thesis admits no shades of gray, thus, it is seen to be a fairly rigid and inflexible mode of logic. But it's utter simplicity renders it amenable to mathematical manipulation. In fact, George Boole (1815-1864) defined an algebra of binary logic based upon this Aristotelian philosophy, which is the basis of the digital computer of today.

•Rule-Based Systems

Contemporary expert systems fall into several categories: rule-based systems, frame-based systems, and blackboard systems. Rule-based systems utilize a knowledge base that consists of rules and facts. All the rules are of the form "if A, then B," where A is a single fact or the conjunction of several facts, and B is another fact. For example, consider a search of a historical data base for ancient personalities: If (1) the person is of Mediterranean origin, and (2) he is known to be Greek, and (3) he is said to be philosopher, then he can be identified as Aristotle with some prescribed degree of certainty. In this example, we narrowed our search from the Mediterranean, to Greece, to the set of philosophers, to Aristotle.

•Frame-Based Systems

In frame-based systems, knowledge is stored in frames. These frames are essentially data bases and can be very large. They store rules, calculate procedures, and function as pointers to other frames or information. Frames are hierarchically ordered and "inherit" properties from the frames residing above them. Frame-based systems are useful in handling complex situations involving objects and classes of objects.

•Blackboard Systems

Blackboard-style expert systems take their knowledge from several knowledge sources which can be embodied as separate data bases. The parent system calls in the information it needs with an appropriate subroutine. Each knowledge system essentially acts as a separate expert system, and conclusions are sent to a shared data base (the blackboard) that is accessible to any data base. The knowledge sources can either be rule-based or frame-based. In this manner, blackboard systems permit one knowledge source to make use of the conclusions drawn by another knowledge source, even if it is structured differently. They are used most commonly when events arrive at the system sporadically and many levels of knowledge are required.

All of the above systems apply the rules by using three common methods: the data driven method (also known as forward chaining), the goal driven method (also known as backward chaining), and the mixed method. All rule-based systems utilize two data bases; one that contains permanent rules and facts, and a second that stores conclusions and facts provided by the user.

A data-driven expert system may appear to function by trial-and-error, seemingly conducting an interrogation of unrelated questions. The logic of the approach is as follows: if there is a rule "if A, then B" and you know the fact A, this leads you to B; then if you find a rule "if B, then C," this leads you to C, and so on--and so on--and so on. Thus, a data-driven search may seem aimless as it wanders on its search through its rules-and-facts domain.

On the other hand, a goal-driven system considers only rules that relate to a specific goal. The system attempts to prove a goal (the "then" portion) by chaining backward through its rules ("then B, if A"). If B is the goal, the system looks for a rule that has B as its conclusion. As the system searches backward through its rules, it eventually reaches a rule that has as its "if" portion a fact that is not the conclusion of any other rule. At that point the system queries the user for additional relevant facts. If the user doesn't offer additional information, the system may be programmed to explore other lines of reasoning. Thus, a goal-driven system functions more like a human expert than a data-driven system. It picks a hypothesis (the goal) and then attempts to prove it, rather than vice versa. The primary advantage of goal driven systems is that they don't seek information or apply rules that are unrelated to their ultimate goal. The disadvantage is that users have little interaction with the process and cannot exert meaningful guidance. This can make goal-driven control unacceptable when a rapid response is desired.

•Mixed Systems

The mixed method utilizes both data-driven and goal-driven control. It permits users the option to volunteer information that the system uses in pursuit of its goals. However, whichever methods are applied, an expert system must contain a strategy for uniquely choosing rules when more than one applies. To achieve this end, some systems have rules regarding rules. For example, a program might contain a rule specifying, "If parameter B exceeds index (I), rules 21 through 25 do not apply."

Despite all of the seeming accomplishments of modern-day expert systems, there are still significant problems to be faced in their use. Expert systems are fundamentally "diagnostic" systems. But expert systems, unlike human experts, have no conception of the needs, wants, limitations, and behavior of their questioners. This severely limits the ability of knowledge based systems to interact intelligently with a user. In the interaction, it often becomes important for the user to recognize the limits of the domain of expertise of the system. Expert systems can be useful; they can also be hazardous. The most obvious danger would be for a user to rely upon a program that wasn't truly expert; one that had signicant deficiencies in its knowledge base and could not recognize a hazardous situation. One should realize that computers and computer programs don't really have any judgement. They don't know what's unreasonable.

Deterministic Methods

The current wave of popularity of expert systems has completely overshadowed the long-standing efforts of the servo controls industry. The theory and practice of the control of engineering systems has long been a major concern of both academicians and practicing engineers. The development of the digital computer and the increasing trend toward direct digital control has led to a demise in the importance of linear systems analysis and network graphing methods. The exceptional reliability and ease of design of digital circuitry has caused the emphasis to shift from linear controls and circuitry to computer managed controls and techniques (or automation). This fact has been the major influence in the advancement and acceptance of Computer Science as a recognized engineering discipline. And this fact has resulted in a major shift in the thrust of engineering development to one of software design and control of engineering systems. AI is an obvious and predictable result of these conditions. Few people seem to remember that *automation* grew out of the old concepts of linear systems theory, long before the term AI was coined by the computer scientists. Automation (or cybernetics) was born in the deterministic concepts of optimal, adaptive, and learning control systems theory.

•Automatic Control Theory

Engineering is a practice concerned with understanding and controlling the resources and forces of nature for the benefit of mankind. The twin goals of understanding and control are complementary; since, in order to control more effectively, the system under control must be understood completely and be mathematically modelled. Control systems engineering is based on the foundations of feedback theory and linear systems analysis, and it integrates the mathematical concepts of network and communications theory. For this reason, control systems engineering is a

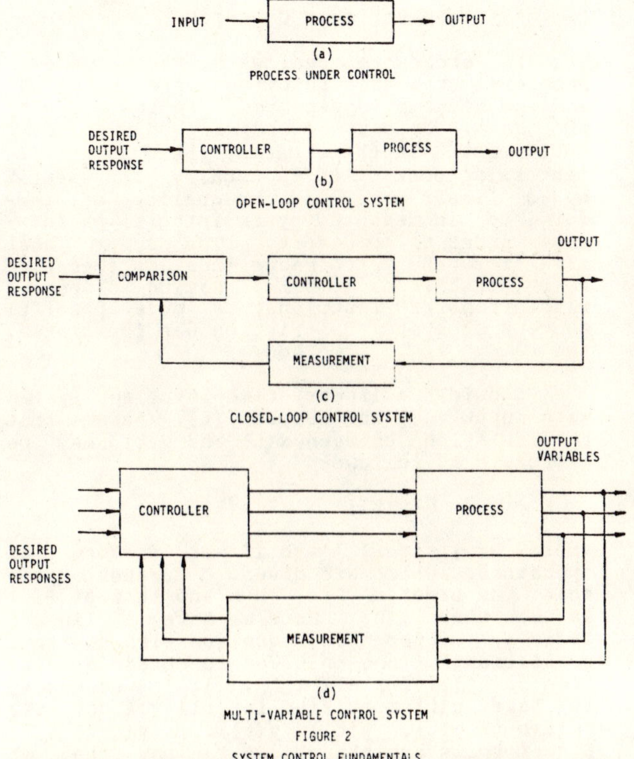

FIGURE 2
SYSTEM CONTROL FUNDAMENTALS

practice not limited to any particular engineering discipline, but is equally applicable to aeronautical, chemical, mechanical, civil or electrical engineering. Furthermore, the principles are as easily applicable to the understanding of the dynamics of business, social, political, and economic systems as well as to physical systems.

A *control system* is a network of components forming a system configuration which will provide a desired response to a specific stimulus. The basis for analysis of a controlled system is the foundation provided by linear system theory which assumes a cause-effect relationship for each component of a system. Therefore, a component (or process) to be controlled can be represented by a block as shown in Figure 2(a). The input-output relation represents the cause and effect relationship of the process; which in turn, represents a processing of the input signal variable to provide an output signal variable, often with a power amplification. An *open-loop* control system utilizes a controller (or control actuator) to obtain the desired response as shown in Figure 2(b). However, in contrast to an open-loop control system, a *closed-loop* control system utilizes an additional measurement of the actual output to effect control of the output with the desired system response. A simple *closed-loop feedback* control system is shown in Figure 2(c). A standard definition of a feedback control system is as follows:

A feedback control system is one which tends to maintain a prescribed relationship of input and output system variables to one another by com-

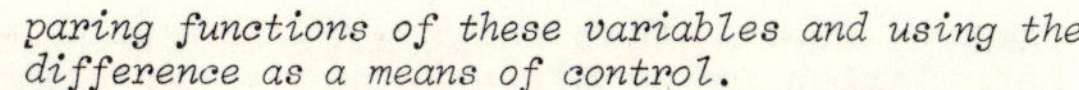

paring functions of these variables and using the difference as a means of control.

The concept of feedback has been the foundation of control system design for many years. Due to the increasing complexity of the systems to be controlled and the interest in achieving optimum performance, the importance of control system engineering has grown immeasurably. Furthermore, as the systems have become more complex, the inter-relationship of many controlled variables has been proved necessary to be considered in the control scheme. A block diagram depicting a *multi-variable* control system is shown in Figure 2(d).

The use of feedback for system control has shown a fascinating history. The first examples of feedback control are found in the development of float regulator mechanisms in Greece in the period 0 to 300 B.C.. The waterclock of Ktesibios utilized a float regulator. An oil lamp devised by Philon in approximately 250 B.C. utilized a float regulator to maintain a constant level of fuel oil. Heron of Alexandria, who lived in the first century A.D., published a book entitled *Pneumatica*, which outlined several forms of water level mechanisms utilizing float regulators. The first feedback system to be invented in modern Europe was the temperature regulator of Cornelius Drebbel (1572-1633) of Holland. Denis Papin (1647-1712) invented the first pressure regulator for steam boilers in 1681. The regulator bore great similitude to a modern pressure-cooker valve. The first automatic feedback controller used in an industrial process is conceded by most to be James Watt's flyball governor, developed in 1769 for the purpose of controlling the speed of his steam engine.

The period preceding 1868 was characterized by the development of automatic control systems by intuitive invention. Efforts to improve their performance characteristics led to worsening transient oscillations and eventually to violently unstable systems. It thus became imperative to develop a mathematical theory of automatic control. James Clerk Maxwell formulated a mathematical theory which could be related to control theory and utilized a model of a governor described by differential equations. Maxwell's study was concerned with the effect of various system parameters upon the system performance. During the same period, I.A. Vyshnegradskii developed a similar theory of regulators in Russia. Control theory and practice developed in the western world in a manner different than in Eastern Europe and the Soviet Union. A major influence in the use of feedback in the United States was the development of the telephone system and electronic feedback amplifiers by Bode, Nyquist, and Black at the Bell Telephone Laboratories. The frequency domain was the primary space used to describe the operation of the feedback amplifiers in terms of bandwidth and other frequency variables. In contrast, the eminent mathematicians and applied mechanics enthusiasts in Russia inspired and dominated the field of control theory, with a Russian theory that tended to utilize a time-domain formulation using differential equations.

A considerable thrust to the theory and practice of automatic control occurred during World War II, when it became necessary to design and fabricate autopilots, gun-pointers, antenna controls, and other military systems based on the feedback control approach. The complexity and expected performance characteristics of these military systems necessitated an extension of the available control techniques and fostered an interest in the development of new methods and insights. Prior to 1940, the design of control systems was practiced more as an art, than a science. In most cases, it was a black art, involving a trial-and-error approach. During the decade of the 40's, analytical methods increased in number and span of utility, and control engineering finally became an engineering discipline in it's own right. Frequency-domain techniques continued to dominate the controls field following World War II with the increased emphasis upon the Laplace transform and the complex frequency plane (the S-Plane). During the 50's, the emphasis in control theory was on the development and use of S-Plane methods, particularly, Evans' root locus approach. Furthermore, the utilization of both analog and digital computers became possible during the 50's. These new control elements possessed an ability to compute rapidly and accurately, functions that were not readily available beforehand. And with the advent of Sputnik and the space age, a new thrust was imparted to controls engineering. It became necessary to design complex control systems for missiles, space probes, and satellites; and the necessity of minimizing weight and highly accurate control has spawned the concepts of optimal and adaptive control. Because of these requirements, the time-domain methods of Liapunov, Minorsky, and others were met and reviewed with great interest. Thereafter, new theories of optimal control were developed by L.S. Pontryagin in Russia and R. Bellman in the United States. It now appears that control engineering must consider both the time-domain and the frequency-domain approaches simultaneously in the synthesis and analysis of control systems.

Control systems engineering is primarily concerned with the design of goal-oriented systems. Therefore, the interests of goal-oriented policies has led to a hierarchy of goal-oriented control systems. But, modern control theory is more concerned with systems which display self-organizing, adaptive, learning, and optimal qualities. In recent years, control theory has been involved in a period of evolutionary development in which significant advances have been made. For problems involving non-linear, time-varying systems, the frequency domain approach became too restrictive; primarily because this approach always resulted in a linearized version of the original system. And if the time-varying parameters varied slowly with respect to a normal duty cycle of the system, these variations were normally neglected.

• State Variable Concepts

In order to deal with these general problems, transfer functions were abandoned; and the systems represented, in the majority of cases, by sets of ordinary differential equations. In these cases, the dynamic element being controlled is usually represented by a linear model and the quality of performance is measured by an integral performance index of some sort. The system that minimizes the performance index is then said to be "best", or *optimal*. Having dispensed with transfer functions, a more powerful concept of system description was formulated, the *state variable* method.

Consider a linear, time-invariant system with input $x(t)$ and output $y(t)$. Assume that the relation between $x(t)$ and $y(t)$ may be expressed as follows:

$$L[y(t)] = P[x(t)],$$

where $L[\ ]$ and $P[\ ]$ are linear differential operators, which are given. Further assume that the order of $L[\ ]$ is n and that of $P[\ ]$ is m, where $m \leq n$. Here we have a linear, ordinary differential equation with constant coefficients, and in order to obtain a solution for $y(t)$ for all $t \geq t_o$, it is necessary to have $x(t)$ specified for all $t > t$ and the values of $y(t_o)$, $y'(t_o)$, $y''(t_o)$,..., $y(n-1)(t_o)$. Since t is arbitrary, we can say that at each instant of time, the response of the system defined by $L[\ ]$ and $P[\ ]$ is uniquely determined by the input and the initial conditions. This viewpoint is entirely equivalent to that obtained by considering the output of the system to be given by the convolution integral

$$y(t) = \int_{-\infty}^{t} h(t-\tau)x(\tau)d\tau.$$

Therefore, we can say that the output at each instant of time is determined by the input considered over all past time. To obtain $y(t)$ for all $t \geq t_o$, it is only necessary to know the impulse response of the system, $h(t)$, and the input for all time. The two differing viewpoints become the same when one considers that the information concerning the input for all previous time is contained in the initial conditions. The viewpoint that we wish to extend here is that the response may be specified at any instant of time in terms of the the initial conditions and $L[\ ]$ and $P[\ ]$. These initial conditions can be specified by n numbers, whose definition will define the future behavior of the system. The magnitudes of these numbers will specify the *state* of the system and the variables used to represent these numbers at each instant of time will be called *state variables*. Knowledge of the state variables at any instant of time will specify the state of the system at that time. In general, it could be said that the state of a system can be defined as a minimal set of numbers given at $t = t_o$ from which, with knowledge of the input, $x(t)$, for $t \geq t_o$ the response of the system is uniquely defined. It may be apparent that

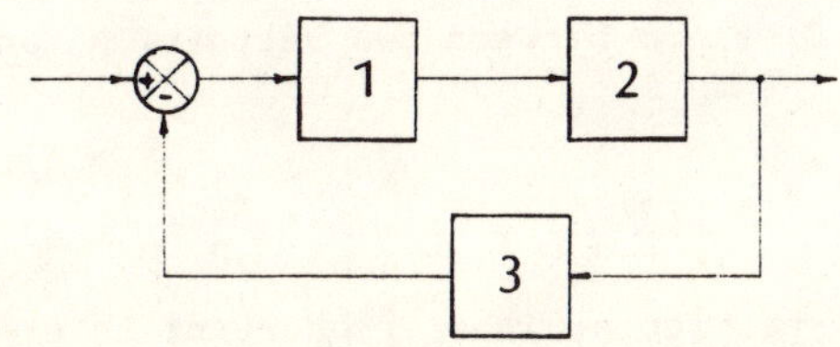

FIGURE 4

AN EXPLICIT FEEDBACK CONTROL SYSTEM

any set of numbers satisfying this definition may be regarded as state variables.

In general, the equations of motion of a typical linear state variable in state space may be written as:

$$\frac{dx_i}{dt} = \sum_{j=1}^{n} a_{ij}x_j + \sum_{j=1}^{m} b_{ij}u_{ij} \quad (i = 1,2,\ldots,n),\ m \leq n$$

$$y_i = \sum_{j=1}^{n} c_{ij}x_j \quad (i = 1,2,\ldots,r),\ r \leq n$$

To define a system of n state variables, with m inputs, and r outputs, this becomes in vector matrix notation

$$\{\dot{x}\} = [A]\{x\} + [B]\{u\}$$
$$\{y\} = (C)\{x\}, \text{ where}$$

$$\{x\} = \begin{bmatrix} x_1 \\ x_2 \\ \cdot \\ \cdot \\ \cdot \\ x_n \end{bmatrix} \quad \{u\} = \begin{bmatrix} u_1 \\ u_2 \\ \cdot \\ \cdot \\ \cdot \\ u_m \end{bmatrix} \quad \{y\} = \begin{bmatrix} y_1 \\ y_2 \\ \cdot \\ \cdot \\ \cdot \\ y_r \end{bmatrix}$$

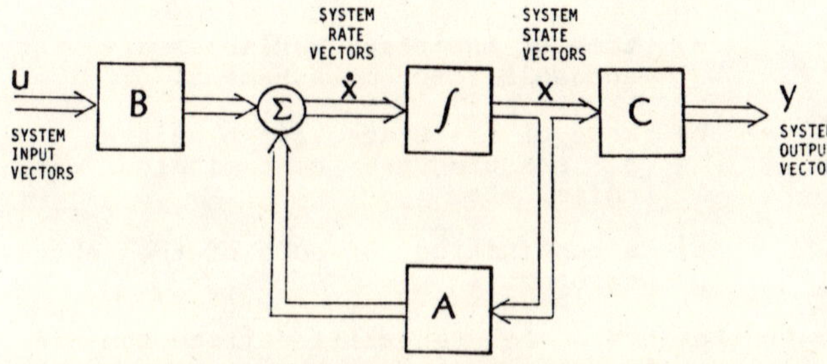

FIGURE 3

A GENERAL MATRIX BLOCK DIAGRAM FOR LINEAR SYSTEMS

Here,
x	=	a state variables
y	=	a system output
u	=	a system input
$\{x\}$	=	the system state vector (n)
$\{y\}$	=	the system output vector (r)
$\{u\}$	=	the system input vector (m)
$\{\dot{x}\}$	=	the system rate vector (n)
[A]	=	the system matrix (nxn)
[B]	=	the input matrix (nxm)
(C)	=	the output matrix (rxn)

The relations stated above are shown in a block diagram analog in Figure 3. In these equations, the x's are the state variables, the u's are the inputs to the system *plant*, and the y's are the *plant* outputs. This statement requires that we define the term "plant". Consider the block diagram of

Figure 4. Let us assume that Blocks 1 and 3 represent what may be termed controllers, actuators, transducers, etc. Block 2 will represent the dynamic element begin controlled and will be called the "plant". The "system" is the entire closed-loop configuration and is not specified until Blocks 1 and 3 or their equivalent is specified. In writing equations, we must be careful to distinguish plant equations from system equations.

Perhaps we should add a brief discussion here on the geometric interpretation that may be attached to the state equations previously discussed. The original sets of linear ordinary differential equations were reduced to a single nth order differential equation. We have succeeded in writing this equation as n first order differential equations. The vector

$$\underline{x}(t) = \begin{bmatrix} x(t) \\ x(t) \\ \cdot \\ \cdot \\ \cdot \\ x(t) \end{bmatrix} = \underline{x} \quad \text{(The underscore indicates the variable is a vector)}$$

has been called the state vector. It is possible (and highly desirable) to attach a geometric meaning to this vector. We may say that $\{x(t_1)\}$, (the value of the state vector, or the state of the system at $t=t_1$) represents a point in a space of n dimensions (n-space). The coordinate system is composed of the mutually orthogonal axes x_1, x_2, ..., x_n. Therefore,

$$\{x(t_1)\} = x_1(t_1)\underline{i} + x_2(t_1)\underline{j} + x_3(t_1)\underline{k} + \ldots,$$

Where $\underline{i}$, $\underline{j}$, $\underline{k}$, ... are unit vectors directed along the coordinate axes. It is fairly easy to visualize systems of order bounded by n=3. However, when n exceeds 3, we merely generalize the concepts and try not to worry about visualizing the space. Since x(t) is a function of time, the motion described by x in he state space may be considered to be a continuous curve or *trajectory*. As t increases from t_o, the state of the system moves along a trajectory beginning at x(t). In visualizing this motion, you should think in terms of a linear space upon which a *norm* and an *inner product* are defined. A norm is simply a measure of distance within the space, and the Euclidean length of a vector x is defined by

$$\|\underline{x}\| = (x_1^2 + x_2^2 + \ldots\ldots + x_n^2)^{1/2}.$$

The distance between two vectors, x and y, is

$$\|\underline{x}-\underline{y}\| = \left[(x_1-y_1)^2 + (x_2-y_2)^2 + \ldots (x_n-y_n)^2\right]^{1/2}$$

The particular inner product of interest in this arena is the scalar or dot product and is defined by

$$\langle \underline{x},\underline{y} \rangle = x_1 y_1 + x_2 y_2 + \ldots + x_n y_n .$$

Notice that

$$\langle \underline{x},\underline{x} \rangle = x_1^2 + x_2^2 + \ldots + x_n^2 = \|\underline{x}\|^2 .$$

The linear spaces with which we normally deal are called Euclidean spaces because they conform to Euclidean geometries.

•Optimal Control Theory

Now that we are thoroughly familiar with the state variable concept, let us advance our concerns to those of optimal control, adaptive control, and learning systems. In the past, the design of control systems has relied heavily upon such empirical methods as Nyquist diagrams, Bode plots, root loci, etc.. These methods employed linearized models and provided the designer with qualitative information regarding the effect of the controller (or equalizer) on the response of the system. Several attempts focused on the possibility of using methods borrowed from the calculus of variations. One of these attempts was called *optimal control theory*.

Briefly, optimal control theory is concerned with a variational formulation of automatic control problems and the attempt to solve the resulting problem using methods from the calculus of variations. The number of significant engineering control problems solved to date using optimal control theory is relatively small. However, the methods hold much promise and is generating considerable interest on the part of practicing control engineers. In the presentation shown below, we will rely upon the method of dynamic programming developed by Bellman rather than the more rigorous calculus of variations. This approach permits a greater simplicity in the derivation with only a slight sacrifice in rigor.

Let us consider a typical optimal control problem. The plant (or object to be controlled) is described by a vector differential equation of the form

$$\dot{\underline{x}} = \underline{f}(t,\underline{x},\underline{u}),$$

where, $\underline{x}$ is an n-vector, $(x_1, x_2, \ldots, x_n)$ called the state vector of the plant; and the components x_i, $i = 1,2,\ldots,n$ are called the state variables. $\underline{x}(t)$ is the state of the plant at time t. $\underline{u}$ is an m-vector, $(u_1, u_2, \ldots, u_m)$ and is called the control vector of the plant. The components, $u_i(t)$, $i=1,2,\ldots,$ m are called control functions. $\underline{f}$ is an n-vector, $(f_1, f_2, \ldots f_n)$. The f_i, $i = 1,2,\ldots,n$ are assumed to possess continuous first partial derivatives with respect to all their arguments. The control functions may be either unconstrained or may be required to remain within an allowable range of values; typically $\underline{u}(t)$ may be required to satisfy the inequality

$$k_{1i} \leq u_i(t) \leq k_{2i} .$$

The general constraint on u(t) can be symbolically noted by $\underline{u} \epsilon \Omega$, where Ω is a suitably defined set. We can now summarize the situation with the statement that if the state $\underline{x}$, is given at some initial point in time in the form

$$\underline{x}(t_o) = \underline{c}_o$$

and the control $u(\tau)$ is given for $t_o \leq \tau \leq t$, then the state, $\underline{x}(t)$ is uniquely determined.

The present theory assumes that:

1) the plant is completely controllable, a condition that is fairly difficult to establish in the case of non-linear, time-varying plants;

2) all the state variables are available for measurement (the requirement of observability); and

3) that disturbances of any kind are negligible (the plant is stable).

Such assumptions can seldom be justified in practice, because either:

a) some of the state variables are not accessible for measurement, or

b) some of the state variables that can be measured are contaminated with noise, etc.

c) a combination of both of the above reasons.

Nonetheless, one can still utilize the results of optimal control theory by either building the best possible estimator in the case of noise-corrupted state variables; or obtaining by the use of linear filters, the best possible approximate value of those state variables that cannot be measured. The basic concepts of controllability and observability will be expanded upon here for they are truly crucial to the existence of an optimum solution, even moreso than stability.

Recent investigations in the theory of optimum control has established conditions which must be satisfied in order for the given problem to possess a solution. One of the more important properties that must be satisfied is what has been called controllability. an additional property that must be satisfied is that the plant be completely observable. Intuitively, controllability is a property of a system which guarantees that any initial state can be transferred to any final state in a finite time interval. In particular, if a certain system is said to be completely controllable with respect to the origin, this means that any initial state can be translated to the origin in finite time. It should be noted that a key expression here is _finite time_. A plant may be asymptotically stable and yet not be completely controllable. The controllability concept implies that it is possible to influence each state variable by means of an input signal.

Observability is an expression of our ability to determine what each state variable is doing, based upon measurements made on the system output. Specifically, if the plant output, $y(t)$, is given for $t_o \leq t \leq T < \infty$, then the plant is completely observable if it is possible to determine the initial state $\underline{x}(t_o)$. It should be noted that the ability to determine $\underline{x}(t_o)$ based upon a finite time measurement implies that $\underline{x}(t)$ is known for all $t \geq T$, and indeed for all $t \geq t_o$. This condition is known as the "identification" problem. Perhaps it is now obvious that if each state variable is available for measurement, the plant is completely observable.

In addition to the facts stated above, we now make the following assertion:

"The concept of n-dimensional vector spaces can be utilized in the field of automatic control to develop optimal control systems by _maximizing_ the vector dot product of the control system output vector with the control command input vector. Similarly, some modern filtering techniques have been based on _minimizing_ the dot product of the filter output vector with the input noise vector. Some optimally filtered control systems seek a combination of both."

The assertion is unproven here, but it can be looked upon as a statement summarizing the fundamental mechanics of optimal controls design. The philosophical basis of the theory stems from the mathematical programming (linear, non-linear, dynamic) methods developed by researchers in the field of operations research. It all hinges upon the _Principle of Optimality_, which is attributed to R. Bellman.

The Principle of Optimality

An optimal policy has the property that whatever the initial state and initial decisions are, the remaining decisions must constitute an optimal policy with regard to the state resulting from the first decision.

The principle may be viewed as a means for obtaining an optimal "policy" for a "multi-stage decision process". Therefore, to apply the principle of optimality to the automatic controls problem, the problem must be reformulated as a multi-stage decision process. To do so, we determine the admissible inputs which will determine a desired output and which, in so doing, will minimize (optimize) a chosen performance index. If such a solution actually exists, we can call it an optimal control solution.

•Adaptive Control Theory

In the broadest sense, any system in which a parameter is altered to compensate for a degradation in performance which is brought about by a change in its environment could be called adaptive. In this sense, there must literally be thousands of examples of adaptive control systems in use today. Every automobile with an automatic choke must be considered to be adaptive. But suppose that by clever design, the operating range of a system can be broadened to accommodate the expected variation of the environment without external parameter adjustment. Could this be termed adaptive? If so, there are millions of adaptive systems in successful operation today.

If we can call these systems "adaptive", we then have a basis to define a new, but yet un-named, class of control systems. However, although we acknowledge that such systems are excellent design solutions to specific problems, we prefer to reserve the word "adaptive" for a class of control system we have as yet to describe. An adaptive control system differs philosophically from a conventional closed-loop control system in the same as closed-loop control differs from open-loop control. If closed-loop control is a "concept", then so is adaptive control. If conventional feedback control has objective reality, then so does adaptive control. Sometimes we have difficulty in deciding whether "feedback" is a proper concept. Borderline cases plague us. The adaptive concept is as controversial and non-objective as the original concept of feedback itself, but _not_ more so.

The control engineer is often faced with the following problem. Given a portion of the system (called the "plant") which may not be altered by the designer, and given certain performance specifications in conjunction with other constraints (e.g., maximum forces or accelerations), he must find a satisfactory controller to manage the plant. If the input of the controller is sensitive only to the system input, the resulting system is open-loop. This is a perfectly satisfactory approach, provided the system and its environment and the expected inputs are completely defined beforehand. On the other hand, if these elements are not completely known, the controller should also be a function of the plant output; and the system is then said to be closed-loop. In order to successfully design the closed-loop controller, the engineer must be cognizant of the plant characteristics and how they vary under environmental changes. Conventional control system design will fail if these conditions are not present.

The problem of incompletely defined plant dynamics have prompted a number of imaginative solutions. The suggested solutions compensated for anticipated changes from an initially well-defined plant. But even this approach is difficult (if not impossible), if the environment cannot be sensed directly by the system, but only by noting a degradation in plant performance; or if little (or nothing) is initially known about the plant or the environmnt. In these cases, it might be advisable to organize the overall system in such a manner that it performs certain of the functions conventionally exercised by the designer himself during the design phase. That is, the controller may be called upon to compute, or in some manner _identify_ the characteristics of the plant while the system is in normal operation. The controller must then make a _decision_ as to the manner in which the available system parameters should be adjusted so as to improve the operation with respect to a previously defined index of performance. Finally, certain signals or parameters must undergo _modification_ to accomplish this result. These three functions (identification, decision, and modification), when accomplished in a closed-loop manner by the system itself, constitute the essence of adaptive control.

From a philosophical point of view, the most distinctive aspect of an adaptive control system is the identification feature. The most severe restriction placed on current identification schemes is that the plant is linear and only slowly time variant. Specific schemes may be classified as to whether or not "a priori" knowledge of the plant is needed for successful operation. It is assumed, of course, that the method must perform in real-time; and while the plant is subject to normal control inputs. An identification scheme is called complete if the entire weighting function of the assumed linear plant is the objective, and partial if only certain lesser characteristics are sought.

The second of the three elements basic to adaptive control is the decision function. This problem deals with the development and specification of analytical methods by which system performance can be evaluated and from which a strategy for achieving optimization can be evolved. The most common method of system evaluation is the use of an index of performance (IP). This is a functional relationship involving system characteristics in such a manner that relative performance may be determined by it. The strategy of optimization is identical to that previously considered for optimal systems. It entails an organized search of an abstract space called the adaptive space, which consists of a mapping of the parameters available for adjustment and in which contours of constant IP value exist. The IP is the criterion by which system quality is measured. We can divide presently used criteria into the following three categories: First, conventional system specifications which involve specific measurements, such as phase margin or rise time; second, general criteria which attempt to sum up the overall performance (these are generally expressed as some integral function or error); and third, compound criteria. Based on some functional relationship of system error and other requirements such as the cost of control. The decision problem involves the design of a logical means by which the system evaluates its current condition with respect to the optimum

(by the use of information gained from the solution of the identification problem) and arrives at a plan of action for driving toward the optimum.

The third and final step in the adaptive process is the actual adjustment of the system so as to optimize it's performance. The general process might be called _control signal modification_, since the object of the operation is to modify the error (or control) signal to achieve optimum performance. There are two ways to accomplish this. The first approach is called _parameter-adjustment_. It is a direct extension of conventional procedures in which a fixed compensator is designed to optimize some aspect of system performance for fixed conditions. In the adaptive system, the parameters of the compensator are adjusted by the system itself so as to maintain optimum performance under varying conditions. However, a more general (and more powerful) approach is _control-signal synthesis_. Here it is recognized that adaptive control requires making adjustments to input signals which drive the plant in an optimum manner, and that adjusting the parameters of a series equalizer is an artificial limitation due only to tradition.

•Learning Control Theory

It can be said that an adaptive control system exhibits a crude form of learning in its self adjustment toward an established goal. However, let us define learning in a somewhat more formal manner. A learning system is a higher form of control, as is an adaptive system. However, the learning system differs from the adaptive system, not only in function, but also in area of application. The adaptive system is designed to modify itself in the face of a new environment so as to optimize its performance. On the other hand, a learning system is designed to recognize familiar features and patterns and then to react in an optimum manner, as prescribed by its' past experience or learned behavior. In the former, emphasis is placed on reacting to a new situation, while in the latter, the emphasis is on remembering and recognizing previously encountered conditions. Not only will a learning system require more extensive logic than an adaptive control system, but inevitably it will be limited in its function by what must inevitably be a limited memory capacity. Because the learning system by-passes the complete identification procedure that is so necessary in adaptive control; or stated differently, because for each iteration it has more "a priori" information at its command, it can in theory react more rapidly. On the other hand, in the same amount of time, the learning system should approach the optimum performance more closely by properly utilizing its past experience.

Those concepts of learning which appear to be of immediate use in constructing superior control systems seem to be limited to the following three: pattern recognition, the simple reinforcement model, and the idea of a learning curve. Pattern recognition has been defined as the mapping of a large, detailed and complete description of an event into a simple description. In other words, the essential characteristics are extracted from a mass of data which will not be needed to recognize the event when it occurs again. On the other hand, a reinforcement learning process is one in which certain aspects of the behavior of a system are caused to become more (or less) prominent as time unfolds as a consequence of the application of a "reinforcement operator", T. This operator affects only those aspects of behavior for which instances have actually occurred recently. The procedure is an analogy paralleling the "carrot" and "stick" philosophy. And finally, a learning curve is a plot of index of performance (or adaptation time) of the system versus the number of trials for the same environmental situation. The system will be said to learn if the curve will monotonically approach the optimum value of the parameter.

Probabilistic and Statistical Methods

In applying the methods described above, we must respect the fallibility of the best of "logical" and "deterministic" systems. Although their expected responses are quantified by computation, their observed responses will be noted to vary considerably from those values. The realities of the real world cannot be dispensed with by mere computation. Statistical design and optimization theory is fast becoming an important branch of modern automatic control theory, and of AI. The conventional techniques of classical design in many disciplines hinges on the use of certain deterministic (and sometimes artificial) inputs, such as sine waves, impulse and step functions. It is often assumed (sometimes without justification) that if systems behave well to these inputs, their proper operation is assured for all inputs. However, in almost all applications, the inputs and/or disturbances encountered by the system can never be completely described in a deterministic sense. Dissatisfaction with this state of the art has led to the development of statistical procedures which handle these problems in a more realistic manner. One class of problem in particular for which statistical design methods are applicable are those in which the dynamics of the plant are precisely known. However, the inputs (and disturbances) to the plant are assumed to be describable in a probablistic manner only. Of increased complexity are design problems in which the dynamics of the plant, or some parameters associated with them, as well as the inputs (and disturbances) are known only in a probablistic manner. The latter finds a number of applications in the theory of adaptive processes. The investigation of problems falling within the first category received a great deal of attention due to the works of Wiener and Kalman in connection with optimum filters. The second category of problems has received considerable attention from engineers and researchers in adaptive control theory, particularly in connection with the identification problem.

Heuristic Methods

The final category of engineering methodologies is *heuristics*. Its origin lies in the Greek word "heuriskein". It means to discover, or learn. Heuristic methods or procedures utilize empirical rules of thumb to find solutions or answers. As applied to AI, it sometimes means little more than good old-fashioned guessing. It was considered by Simon and Newell in the mid-50's as an AI approach. They thought that symbol processing (rather than computing) would be the ultimate role for computers. To explore this potential, they began programming computers to prove theorems in symbolic logic. At first they tried to exploit the speed and capacity of the computer with a comprehensive "generate-and-test" algorithm. The computer was successful on a few theorems, but it became obvious that it would require hundreds of years to obtain some of these theorems in this "brute-force" manner. They then studied the behavior of students in proving the same theorems, and found them to be faster than the computer because they made clever guesses instead of rigorously generating and evaluating every possibility.

By incorporating the guessing rules into a computer program (called the Logic Theorist), Simon and Newell discovered that each rule worked for only a few theorems, but that the computers theorem-proving speed increased by orders of magnitude. In this way, Simon and Newell discovered a fundamental principle of AI--determine the manner in which human experts "guess" their way through large complex search spaces and mechanize this guessing technique on a computer. The guessing rules are called "heuristics", and they differ from algorithms in the mathematical sense in that they either produce relatively fast solutions, or none at all.

SUMMARY

In the preceding pages, we have hopefully shown the relevance of present-day expert systems and the contributions of modern control theory. Each is seeking to solve the same problem, but the individual approaches are widely divergent. Expert systems are essentially "diagnostic" systems, depending upon large data bases and symbolic manipulation to effect their purpose. This is the approach chosen by the avowed computer scientist to the emulation of human thought processes. On the other hand, the controls systems engineer has chosen to implant human-like intelligence in physical systems via classical linear mathematics techniques. The use of the digital computer and statistics and probability measure is the common denominator between the two. Before each has reached his common goal, they may well find that each one needs the other. But one thing is highly apparent; the mechanics of human thought is neither simple nor digital in nature. The process of learning will not be easily fathomed.

To return to the original questions posed by the emergence of AI-- "what is intelligence?" and "how can machines be made to exhibit intelligence?" The answers to these questions are not yet at hand. Despite the premature claims of many proponents of expert systems; "a computer program does not really think!!" And despite the best efforts of many highly acclaimed controls systems engineers, the design of optimal, adaptive, and learning systems has yet to be reduced to simply managed procedures. The first blush of success has faded from optimal-adaptive control in much the same manner as complete success has eluded expert systems. But we must not despair. We must dig in for the long, hard pull to the design of useful systems. The trivial problems have been solved and the academic pontificators have left for new areas of confusion. The approach to the solution of the adaptive problem by dividing it into the sub-problems of identification, decision, and modification may yet prove successful. In identification, cross-correlation is being used occasionally, but the fundamental limitation of the identification time is a major drawback. In recent years, attention has been shifted to methods which assume certain "a priori" knowledge of the plant; such as the form of its governing differential equation. The decision and modification problems may be viewed as a "hill-climbing" problem, while the gradient methods continue to be useful. Learn-ing control concepts are not definitively established as yet and considerable research remains to be done.

There are many similitudes between expert systems and controls systems methods. Not the least of these similarities is the establishment of a solution space and the need to search that space. Many disciplines have contributed to both approaches to the AI problem; and these include mathematics, operations research, engineering, and to some degree -- psychology. The terms "adaptive" and "learning" have been used by various persons to describe the behavior of certain inanimate objects, presumably by analogy to the actions of various forms of living organisms.
This analogy may be explicit and express, or it may not physically exist. However, if the engineer is so undisciplined as to abuse this analogy, he may run the risk of ignoring the benefits of almost a century of research on the learning problem. It would seem that proper application of such analogies would aid the efforts of researchers in cybernetics; and then in turn, would illuminate the learning problem in psychology and the behavioral sciences in general.

The difficulties of an engineer attempting to use the psychology of learning are admittedly formidable and must be acknowledged. However, as Woodworth and Marquis pointed out some years ago [22], "No general theory of learning commands the universal approval of psychologists at the present time. Approaching the problem from different sides and by the aid of different experiments they are inclined to emphasize different factors." More recently, George [6] stated, "now it can be seen that there are large possibilities for confusion in our basic understanding of 'learning' and more careful definition is necessary. If one turns to some modern definitions of 'learning' it may be seen that greater clarity has been only partially effected." In the 40's, psychologists began to emphasize the mathematical theories of learning, and since then, certain of the concepts of cybernetics began to be discussed. More recently, a growing interest on the part of psychologists in the concepts of feedback control and mathematical learning theory [15] became evident. The cross fertilization between psychology and automata theory did yield a few devices of real interest to both fields. The Perceptron of Rosenblatt [18] was one notable example.

As a final note, the writer assesses AI as follows. AI (or MI -- machine intelligence) was neither conceived nor developed by the Computer Science community. It was conceived many centuries ago by persons such as Babbage and Lady Byron. In more recent times, it has been revived by Norbert Wiener in his concept of "Cybernetics". Both ideas pre-date McCarthy's landmark meeting of 1956. As for development of the concepts of "adaption" and 'learning', the beginnings of automatic control theory also pre-date the Dartmouth conference. Furthermore, it is well known that the brain, unlike today's digital computer, utilizes a continuous variable format; not a discrete binary format. Also, the brain is highly parallel in function; not serial. At present, this person does not see success in generating cognitive thought as belonging to any specific discipline. What he does see, if we can overcome the vested interests of certain individuals, is an amalgamation of effort on the parts of the computer science, engineering, mathematics, operations research, and (most of all) the psychology disciplines in a concerted effort to create a breakthrough to machine generated cognitive thought.

BIBLIOGRAPHY

1. Alvarez, J. C. 1983.   _A Survey of Artificial Intelligence_. Ford
    Aerospace Corporation.

2. Athans, M., and Falb, P. 1966. _Optimal Control_. McGraw-Hill.

3. Bellman, R. E., and Dreyfus, S. E. 1962. _Applied Dynamic Programming_.
    Princeton University Press.

4. Chen, C. F., and Haas, I. J. 1968. _Elements of Control Systems Analysis:
    Classical and Modern Approaches_. Prentice-Hall.

5. Dorf, R. C. 1967. _Modern Control Systems_. Addison-Wesley.

6. George, F. H. 1962. _The Brain as a Computer_. Addison-Wesley.

7. Gevarter, W. B. 1983. _An Overview of Artificial Intelligence and Robotics_.
    NASA.

8. Ham, M. January 1984. _Playing by the Rules_. PC World.

9. Hawkins, R. D. 1983. _Buffered Receptor, Avionics Integration Network
    (BRAIN) -- A Concept Proposed for Memory Managed Avionics_. NAECON.

10. Hill, J. D. et al. 1964. _Modern Aspects of Automatic Control_. Purdue
    University.

11. Kinnucan, P. January 1984. _Computers That Think Like Experts_. High
    Technology.

12. Lerner, E. J. August 1984. _Why Can't a Computer Be More Like a Brain_.
    High Technology.

13. Manuel T., and Evanczuk, S. November 1983. _Artificial Intelligence_.
    Electronics.

14. Morris, H. M. May 1984. _Robotic Control Systems: More Than Simply
    Collections of Servo Loops_. Control Engineering.

15. Mowrer, O. H. 1960. _Learning Theory and the Symbolic Processes_. John
    Wiley & Sons.

16. Naedel, R. G. August 1984. _Intelligent Associative Memory: An Overview_.
    Defense Science 2001+.

17. Pournelle, J. November 1984. _Artificial Experts_. Popular Computing.

18. Rosenblatt, R. 1962. _A Comparison of Several Perceptron Models_. Spartan
    Books.

19. Schultz, J. B., and Russell, D. M. January 1983. _Weapons That Think_.
    Defense Electronics.

20. Selfridge, O.G. 1959. _Pandemonium: A Paradigm for Learning_. Mechanics
    of Thought Processes, Vol I.

21. Taylor, E. C. July/August 1983. _Artificial Intelligence in the Air-Land
    Battle_. Astronautics and Aeronautics.

22. Woodworth, R. S., and Marquis, D. G. 1947. _Psychology (Fifth Edition)_.
    Henry Hold & Company.

23. Zadeh, L. A. August 1984. _Making Computers Think Like People_. IEEE
    Spectrum.

The distributed intelligence system and aircraft pilotage

LEIGHTON L. SMITH, PH.D.

UNIVERSITY OF CENTRAL FLORIDA
ORLANDO FLORIDA

A Distributed Intelligence System is a complex human-machine system where an artificially intelligent operating component (AIOC) operates the system and the human manages the system. In order for the AIOC and the human to work together synergistically the DIS needs to be carefully designed. There are a wide variety of human-machine systems which would benefit from a DIS application e.g., aircraft pilotage. The pilotage task is discussed in detail.

INTRODUCTION

An artificial intelligence system is a system which at one time was operated by a human and now is operated by a machine. A distributed intelligence system (DIS) is a system which at one time was operated by a human, but is now operated by a machine and managed by a human. This machine has enhanced capability due to application of artificial intelligence (AI).The human is still involved in the activities of the system usually making decisions, planning, and performing other high level functions while the machine portion of the system executes most of the regular operational functions of the system, collects and stores data, and handles routine decision situations which have a known and finite number of options.The DIS is an intermediate stage in the transition from totally manual to totally automated systems.

There are many complex and sophisticated human-machine systems which are beyond the scope and capability of current AI technology yet which are systems which could significantly benefit from the application of some form of machine intelligence. Such systems would benefit from this sort of AI technology application by being able to perform at a higher level in less time with less people.

Therefore the incorporation of some form of artificially intelligent operating component (AIOC) into a complex human-machine system is in many cases a highly desirable system design modification.

SYMBIOSIS

To design a DIS by incorporating an AIOC into certain portions of a human-machine system does not guarantee enhanced system performance with the same or less input. The humans who remain involved with the resulting DIS must be willing and able to work with their new AIOC partner. There are a number of compatibility considerations which must be addressed in the design phase.

Generally humans have to coped with a natural aversion to working with machinery and equipment because these things were inanimate and unintelligent and the humans were able to control, operate, design, and maintain them.

Now with the incorporation of AIOCs into human-machine systems such terms as "user acceptance", "user friendly" and symbiosis are being coined to describe a new burgeoning relationship between humans and machines. A well designed and appropriately applied AIOC can be threatening to a human. The AIOC is incredibly fast, makes no mistakes, does not complain, and never gets tired.

In addition, consider the problem of how a human will communicate with an AIOC. Note that humans communicate with each other generally on a high level but at a low bandwidth. This means that we can exchange very complex information with each other while using very simple means to do so. Whereas machines capable of communicating with other machines usually do so on a low level and at a high bandwidth. This means that only a small amount of simple information can be exchanged at a time. Therefore not only does a DIS design need to address the compatibility of interface but also the compatibility of symbiosis.

APPLICATION EXAMPLE

A typical complex human-machine system is that of aircraft pilotage. A human pilot is the operator of the system. Aircraft pilotage is a time-shared task. There are innumerable subtasks for which the pilot is responsible. In a typical flying maneuver a pilot must
A.   Monitor
    1. Airspeed
    2. Altitude
    3. Vertical velocity
    4. Accelerometer
    5. Tachometer
    6. Exhaust gas temperature (Jet Turbines)
    7. Hydraulic pressure
    8. Oil temperature
    9. Oil pressure
    10. Fuel supply
    11. Emergency indicators
    12. Altitude indicator
    13. Compass
    14. Distance measurement equipment
    15. Visible horizon
    16. Surrounding airspace (for other aircraft... called clearing)
    17. Flight plan
    18. Maps/Charts
    19. Inflight guide
    20. Avionic displays (ILS, HUD)
    21. Computer

B.   Adjust as necessary
    1. Throttles

2. Flight controls (joy stick: ailerons and elevator, and trim)
3. Rudder
4. Radio frequencies (VHF, UHF)
5. Navigation aids (VOR, TACAN)
6. Oxygen equipment
7. Configuration (landing gear, flaps, speed brakes)
8. Visors
9. Cockpit heater/air condition controls
10. Transponder
11. On board computer

C.   Plan maneuver with respect to area to be utilized and energy level to be maintained

D.   Think ahead two, three (or more if possible) maneuvers with respect to area, energy, and fuel supply

E.   Communicate with ground control and other crew members

F.   Implement tactical weapons systems (machine gun, rockets, missiles, bombs).

In an F-16 at a maneuvering speed in the range of 350-500 knots, a pilot can perform a typical maneuver time-sharing among all these factors in about one minute. It is impressive when the assimilation and interaction of information by the pilot is detailed in this manner, that the maneuvers get executed at all. The key to repeated maneuver success in this system is that usually everything in the aircraft/weapons system goes as expected, i.e., nothing goes wrong. When something does go awry maneuver success is always jeopardized and often it is sacrificed. This is because it takes time for the pilot to deal with the problem and most maneuvers do not have much flexibility with respect to time.

Consequently it can be viewed that a human-machine system such as aircraft pilotage is a suitable candidate for application of a DIS. It is clearly feasible that the incorporation of one or more AIOCs into a pilot-aircraft system would enhance maneuver success because the pilot could be freed from some of his/her time-sharing responsibilities and thus have more time available to deal with a variety of problems which might develop.

The key question is which facets of the pilotage task are most appropriate (i.e., effective) applications for AIOCs. The answer to this question is predominantly dependent on the answer to the ancillary question: which responsibilities is the pilot willing to give up?

SYNERGISM

The issues surrounding a DIS application to a complex human-machine system such as pilot-aircraft are so critically delicate that it would seem more prudent not to disturb a reasonably successful performance record. This would, no doubt, be a more than adequate disposition had not such systems recently become more complex and in turn exceedingly more expensive. Greater capital outlay naturally translates into greater cost of substandard performance. Greater complexity places a greater demand on the system controller (the human) and training and experience are not enough to compensate. Hence there is system performance degradation.

Thus some form of assist is needed. This assist must be able to handle certain facets of system operation quickly and properly. It must be reasonably priced. And it must be acceptable to and readily used by the human operator. In fact beyond symbiosis this assist must be able to achieve a synergistic relationship with the human.

This is a tall order but it is potentially more simple that it would seem. The actual effect of incorporating one or

more AIOC into a complex system is not that system performance is enhanced. No, this is a result. The effect which causes the result is that the AIOC application makes the human's job easier. If the decision of where to make an AIOC application were based on where would the human operator's burden would be most effectively eased, the most synergistic result would be achieved.

It may prove that in highly trained individuals there are too many transfer of training difficulties to overcome and thus successful application of DIS is only feasible for human operators who are trained from the basics with the DIS in place. This is a very likely prospect for the problem of aircraft pilotage and DIS. It may very well be that there is no cost effective way to isolate which subtasks can be supported by AIOCs. This is likely due to the fact that it would be difficult to extract reliable data on how pilots regard each subtask as part of the overall task. This difficulty is because the subtasks are first not independent and second are highly interrelated with other subtasks and therefore cannot adequately be analyzed separately.

CRYSTAL BALL

In spite of a large number of explicit and implicit difficulties, there is definitely a need for DIS applications in a variety of spectra. There is aircraft pilotage as already mentioned. There are submarine pilotage, airtraffic controlling, crane cab operation, nuclear power plant operation, and space craft launch control to name a few.

The design of AIOCs and the design of DIS are both in the developmental stage and there are a number of pertinent research issues which would be beneficial to be addressed in the immediate future. To wit:

A. Feedback Formats/Media
Much more needs to be known about how a human can effectively communicate with a machine. Namely the machine must effectively convey important information in such a manner that it can be assimilated in a very short period of time. In turn the machine must be able to garner information from the human in the presentation manner most comfortable for the human to express it and then be able to translate such information into a form that the machine can effectively utilize.

B. Higher order compatibility relationships
To best achieve symbiosis, more research in the realm of human-machine compatibility is necessary. The need is to go beyond the current state-of-the-art in control and display compatibility relationships.

C. Psychophysiology
There is a very close relationship between the physiological aspects of a human-machine system and the psychological aspects. How these interact and how system components affect this relationship are current unknowns which would need to be investigated in detail. Also related in this area are fatigue and stress/anxiety. The advent of a DIS system may enhance performance by reducing the tendency towards fatigue in the human and at the same time degradate performance because the wrong application or the wrong format for the application causes an increase in the level of induced stress/anxiety. Note that the application of machine automation generally is warranted if there is a repetitive task, an unpleasant or objectionable task, and/or a hazardous task. Unfortunately, in complex human-machine systems such as aircraft pilotage, these general guidelines do not prove helpful. Consequently there is a need for formal foundation level empirical research in this area.

REFERENCES

Ashley, Steven "Artificial Intelligence in the Plant."
AMERICAN METAL   MARKET/AMERICAN METAL-
WORKING NEWS, January 10, 1983, pp. 7,8,10,11 &12.

Automation News. "Expert Systems - 'Key' to Production
Future." AUTOMATION NEWS, June 13, 1983, p. 10.

Bellman, Richard. ARTIFICIAL INTELLIGENCE, Boyd &
Fraser Publishing Co., 1978.

Evanczuk, Stephen, Manuel, Tom. "Practical Systems Use
Natural Languages and   Store Human   Expertise."
ELECTRONICS, December 1, 1983 pp. 139-146.

Meyrowitz, A.   "Research Directions in Artificial
Intelligence." NAVAL RESEARCH REVIEWS, Vol. 34, N4,
1982, p.10.

Rose, C., Lefkowitz, I., Kahne, S., "Automatic Control by
Distributed Intelligence." SCIENTIFIC AMERICAN, Vol.
240, N6, June 1979, pp. 78-90.

Shortliffe, E., Duda, R. "Expert Systems Research."
SCIENCE, Vol. 220, April 15, 1983, pp. 261-268.

Smith, L. L. "Visual Flight Simulation-Can We Get There
From Here and Will It Be Worth It?" ALL ABOUT
SIMULATORS, 1984, Proceedings of the SCS Simulators
Conference, 1984 v14, April 1984, pp. 40-42.

Smith, L.L., Connally, R.E. "Artificial Intelligence and An
Adaptive Training Concept" PROCEEDINGS The First
International Symposium of Human Factors In
Organizational Design and Management, August, 1984,
Honolulu, HI, pp. 609-612.

Waltz, D. "Artificial Intelligence: An Assessment of the
State of the Art and Recommendations for Future
Directions." THE AI MAGAZINE, Vol. 4, N3, Fall 1983, pp.
55-67.

The simulation environment concept artificial intelligence perspectives

Kevin D. Reilly, Warren T. Jones, and Pradip Dey
Computer & Information Sciences Department
University of Alabama at Birmingham
Birmingham, AL 35294

ABSTRACT

This paper describes on going research in simulation environments. The project involves both the analysis of conceptual bases and the design and development of environments. Goals include provisions for combined continuous and discrete modeling with extensions to include symbolic modeling capability. The conceptual base entails a distributed systems framework which incorporates strong AI components that is assumed to evolve from its initial experimental version to more advanced ones. I.e., though current design and development focuses on a 3B-Net context with equipment provided by AT&T for this purpose, it is expected that in the future this homogeneous network will evolve into a heterogeneous collection of nodes with advanced and specialized processors supporting the various components of an environment. The initial 3BNet environment is being developed by integrating a number of existing software systems including PSL/PSA, System S, and ASP, along with several AI support tools.

CURRENT STATE OF KNOWLEDGE

With increasing frequency, scientists and engineers are turning to computer-based modeling of systems they hope to understand, build or modify. Many tools have been produced to carry out the modeling process itself, e.g., statistical and numerical analysis packages, simulation languages, and the like. Though important facilitation of the modeling process has been achieved in the past, the present state of the art leaves much to be desired. Current problems with software tools and aids include lack of compatibility, ill-defined capabilities, lack of uniformity, lack of tailorability, lack of support for software evolution and lack of evaluative data [1].

A potential solution to this problem lies in the concept of a "Simulation Environment" (SE). A SE is a collection of tools that are well-integrated and interacting synergistically in support of all phases of the modeling process. A characterization as provided by Henriksen [1] is found in Figure 1.

In this paper, we adopt Henriksen's framework and seek to complement it by:

* Proposing potential architectures, realizable on real machines, one on which we are carrying out explorations, and another to which it might be expected to evolve

* Emphasizing the role of AI techniques in various phases, in current research systems designed to fit within the proposed architectural frameworks

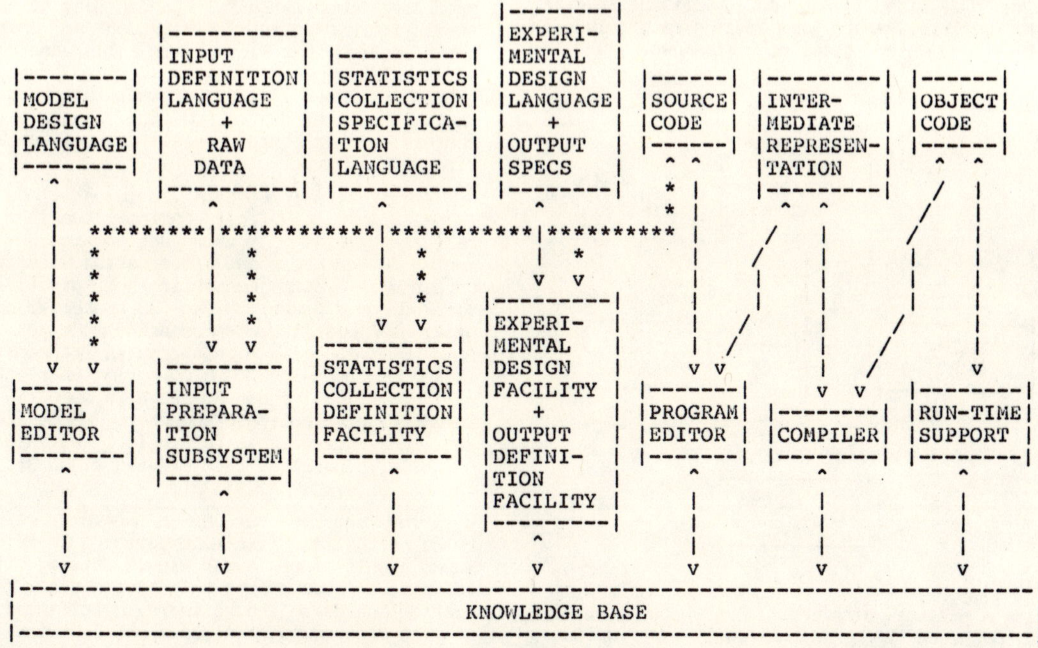

Figure 1: Henriksen's Conception of the Architecture of the SE.

29

For purposes of discussion we will use an aggregation of the Henriksen model into four basic components as shown in Figure 2.

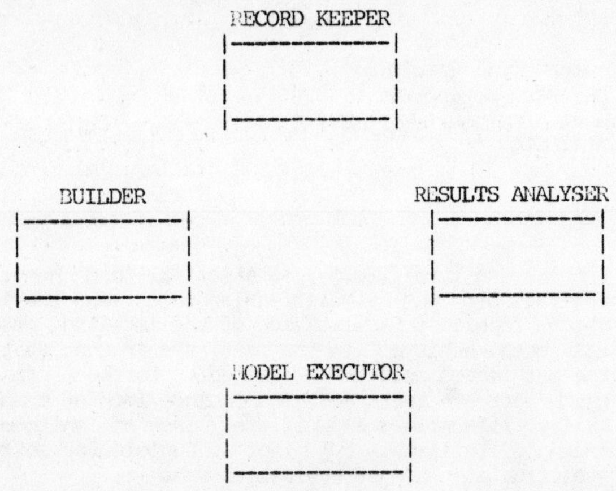

Figure 2: Four primary components of a Simulation Environment.

DISTRIBUTED PROCESSING ARCHITECTURE

On the mechanics and experimental side of the work our focus is on a 3B-Net-based system. A schematic diagram paralleling the elements of Figure 2, and labeled with components in our 3B-Net, is illustrated in Figure 3.

The system, as described in Figure 3, is a loosely coupled net, pitched at a macro-level of processing. One of the reasons for this becomes clear when we discuss Figure 4 below. Other reasons exist. For example, though we have specifically designated processors for building models and for results analysis, there is a kind of interchangeability between the two, based in part on established practice in simulation, i.e., that much modeling takes place in a highly circumscribed statistical framework, whereas

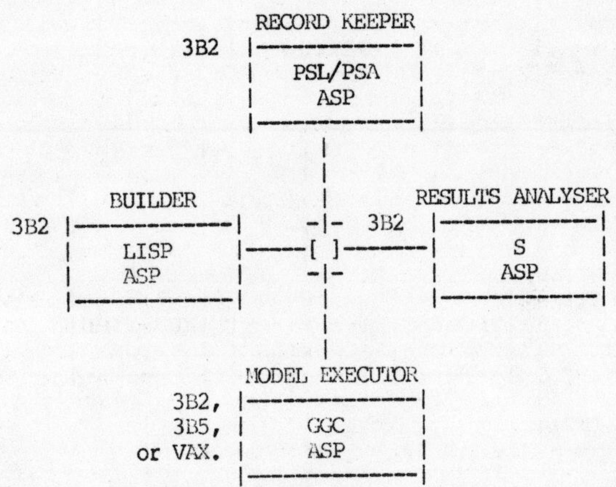

Figure 3: A physical architecture on which to build a SE. The machines to be used and software processors that will run on them are illustrated.

other forms of modeling are more ad hoc, in which accuracy of results is achieved through heavy reliance on replication. The very popular discrete-event system, GPSS, is perhaps the clearest embodiment of the latter philosophy. This split in approach is reflected very broadly throughout the literature on simulation, dividing principally along lines of programming-oriented and mathematically-oriented emphasis. We feel this split is likely to endure, and are designing accordingly, so that it is possible to enter models from either of these types of units and to perform certain forms of output analysis at them.

The presence of the ASP processor on each of the machines is of significance in several regards. The ASP system, which we have developed at UAB, as an extension of an older batch-oriented macro processor, Stage2, developed by Waite [2], is a fully interactive processor with compiler options through links to Lisp (and perhaps, soon, to Prolog) [3]. ASP constructions allow mimicking of Prolog and Lisp styles, among other attractive features such as high portability. The ASP system now runs on the Vax under Berkeley UNIX (in the C programming language) as well as the Data General Eclipse [4] [5] [6] [7] [8], and is in the process of being ported to the UNIX5 C language on the 3B-Net. It is estimated that ASP can be ported to any machine which has a C or Fortran compiler (and actually others such as Snobol!) in less than one-half day's work by one person.

As such, ASP not only fits well into the architecture presented in Figure 3, but in principle allows room for a heterogeneous set of processors. These might be most useful to extend capabilities for the users of the system. Some of the features of software that has already been built in ASP are outlined below. They demonstrate many features that ASP can provide not only in user-system interaction, but also in facilitating communications in a network of the type we are proposing and in providing an interface between heterogeneous modes of processing implied by the software processors of Figure 3. In the user-system interaction purview, ASP-based software, such as a processor dubbed Barrel-F, can act as an analog to shell programming on a system such as UNIX. It is worth noting that Barrel-F is a language which in and of itself has been formally defined using Metamorphosis grammars (i.e. a formal logic semantic grammar). The tight control on the language putatively confers on it better overall control for an application of this type.

Design and (prototype) implementations on 3B-Net is a possibility that is new: the 3B systems were released by AT&T only this year and the UAB's Department of Computer and Information Sciences has just obtained a system consisting of six 3B2 machines (linked on 3B-Net). The 3B system we have secured provides a distributed-processing mode of operation. Some workers believe that such environments will make such a profound impact on computing that the results will dwarf all other advances in computing since the inception of electronic computers about forty years ago. Our edge on this future is illustrated in Figure 4 (again in parallel with Figures 2 and 3).

The fact that our design should gracefully evolve to a situation like that depicted in Figure 4 is of major significance in our view. This architecture, it is noted, presupposes two major systems of "new generation" hardware and software, specifically a system of the supercomputer class and one of the "fifth generation" type (an inference machine) [9]. A supercomputer will likely be purchased by the state of Alabama

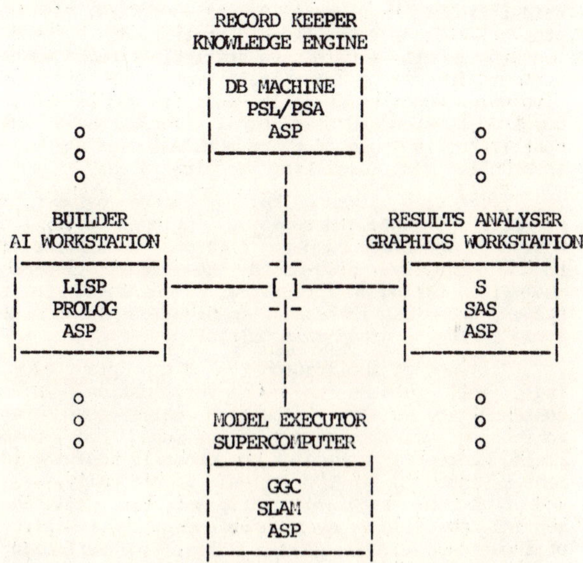

Figure 4: A future architecture for the SE. Multiple versions of the building units and the analysis units are implied by the dots in the figure.

within a year. We shall be content to simulate an inference engine for the near future. It is important to note the network we are proposing satisfies one criterion often associated with supercomputers: that the data rate to them be satisfied by front-end or co-processors.

<u>ARTIFICIAL INTELLIGENCE ELEMENTS</u>

As noted, we intend to emphasize AI techniques. Our experimental work in this area is and will be based on much previous work and covers all the major elements, though more emphasis so far has been placed on the model building tools and the database. The SE concept is not expected to reach full fruition until some time in the 1990s. We expect the considerable emphasis that has been placed on AI tools and techniques to grow. Design work is underway covering several areas that we have not yet done exploratory work. We overview several of our studies later in the paper.

<u>EXPERIMENTAL AND DESIGN STUDIES</u>

We now outline work already accomplished and in various stages of development, using as an organizational principle the primary components we have outlined in Figures 2, 3 and 4.

<u>Building Models</u>

Lisp, long known for its symbolic manipulation prowess and as "the language" of Artificial Intelligence (AI) helps in producing AI components within the SE and with certain advanced forms of modeling [10] [11] [12]. With the advent of "Lisp Machines" (e.g. Symbolics, Lambda, and others) and a full repertoire of floating point arithmetic being a component of most modern Lisp implementations, Lisp has come of age. It is being seriously proposed in some circles as perhaps the best starting point for "Fifth Generation" computing and perhaps even as the "only" language which has

capabilities needed for future comprehensive computer processing.

We have in house a sophisticated version of Lisp for the Vax 11/750 (Franz Lisp), which has the expert system building tool, OPS5, available on call [13]. We have used vanilla Lisp for complex symbolic modeling, with some simple numeric considerations involved [12]. As is not atypical in AI work, we have done elementary statistics assessing the performance of components of the model, e.g., for searching difficulty in terms of the number of "primitive" search activities, for size of a semantic net in terms of its number of nodes as a product of certain kinds of updating, and the like. On the drawing boards, and already investigated in a preliminary fashion, are more sophisticated numerical inputs to the model, reminiscent of some combined continuous and discrete models. Specifically, restrictions are put on entry of new information and extracting of information from the semantic net as determined by random events such as limitations on, e.g., the depth of search or the amount of time allotted for search, and breaking of links within the net. The goals of such studies include modeling some aspects of abnormal behavior, background for which is found in [10] and [12], and includes, in the longer run, the old theme of "reliable computation from unreliable parts".

The production system, OPS5, has been used for model development studies to explore its use as an alternative and complement to table processing in Prolog and ASP, particularly for tables in decision table format (though we treat this latter term broadly, much as does Montalbano [14], to include so-called condition policy maps, action policy maps, and a variety of other extended entry formats). The tasks we have performed relate closely to tightly defined decision table processing in Prolog by Salah [15], so that these tools might be available in a broader context.

In addition to this exploration we have used both general Lisp processing (often in purely functional style) and OPS5 to carry out analyses in statistical data processing. The paper [16] is relevant to explorations of purely functional style in another version of Lisp, coupled to an ASP-based system, but not yet integrated into SE operations. We mention this functionality again, in the Results Analyser section.

Previous work involving natural language processing also possess a number of important feature related to the SE. Specifically, in our memory system models [11], sentence input is accumulated into a form of semantic net. A simple query language allows direct access to the net information, but in recent developments, we have extended the system so that a subset of the information in the net (to include all the information, if need be) can be extracted and reformatted so that it can be processed by Prolog [11]. The importance of this capability includes the fact that it readily extends the query capabilities for anyone knowledgeable in Prolog. On the other hand, it also interfaces with table and graphics processing routines, which with this work constitutes an extremely important potential for merging natural language input with table and graphics data entry, an important problem in AI proper. Finally, it links to the KB, as we relate in the section on the KB below.

Much work done in the context of the ASP processor complements and extends this work. An ASP-based two-dimensional programming system (table-oriented) has been developed and integrated with table processors in a variety of machines and languages [3]. The ability to develop tailored code in ASP contributes greatly to developing processors for various parts of a model, e.g., for block-oriented components or even for stack-queue-limited data structuring for event routines. Such work may prove extremely helpful as the network system becomes more heterogeneous.

Model Executor

The modeling context for the SE is most often associated with numerical computing, with a range of mathematical techniques, e.g., algebra, differential equations, probability and statistics. All of these topics are involved in combined continuous and discrete simulation.

In a recent doctoral dissertation by J. W. Hooper, under our direction and its associated publications [17] [18], we outlined a theoretical and practical approach to this subject matter. Valuable sources for guidance with respect to such theory, distributed simulation, etc., are found in Ziegler's work [19] [20]. These theoretical aspects of simulation, more directed toward building software packages for model expression, though such work also contributes to tasks faced by users in very sophisticated modeling chores.

In the practical end of this work, we developed a prototype combined continuous and discrete simulation processor which incorporates almost all the features of both the very popular discrete event language, GPSS5, and of the the GASP-IV language. Pritsker and Pedgen [21], writing on SLAM, point out that several different modes of processing are possible in this system, to include mixtures of discrete event, transaction, and continuous components. These modes are all available in GGC.

In addition, some other explorations involving symbolic components, with preliminary design for incorporating them into an ASP-based system in intimate communication with GGC, threaten to extend the scope of combined simulation to one including "symbolic events" and "symbolic transactions" as well as the more established forms. Decision making based on manipulating of symbolic information has been increasingly involved as the domain of models expands. We have discussed this matter in recent publications, and add here that the SE concept, because it makes extensive uses of symbolic manipulation, when pursued in a manner we advocate, will also contribute to increased symbolic modeling capability [3] [10] [11] [12] [14] [25].

In student projects at UAB, we have explored a number of issues and evaluated prospects for automation of some tools, e.g., in areas of systems analysis and design, with a range of computations, from Monte Carlo applications [22] to simulation, the "internals" of the GGC language [23], to complex combined-discrete simulation models [24]. Though we view these items primarily in light of model execution possibilities, they represent important ramifications for model building and analysis besides. They also play an important role in the PSL/PSA world view, as we discuss in the KB section below.

Results Analyser

The S graphics and data analysis system represents two emerging and correlated approaches: the heavy use of interactive graphics in dealing with data (whether from the "real" world or from simulation models), and the "data analysis" approach, in which an exploratory mode of operating on data is adjoined to conventional (and sometimes not so conventional) statistics [26] [27] [28].

Among the special attractions of S are these orientations: they make S, in our view, much more attractive than conventional statistical approaches (though modern interactive statistical packages have strength in some of these areas too). The simplicity with which sophisticated data analysis can be done, as compared to standard statistical rigor can be a major contributor to use of these methodologies for highly interactive work, e.g., for stopping rules.

S can also act as a halfway house between the modelers of the two types we mentioned above, i.e., between those hide-bound to rigorous statistics and those equally hide-bound to replication-based analysis. Having an integrated S component in the system, then, can increase communications about models among a diverse group of modelers.

Yet another application that S can play exceedingly well relates to model verification. In some contexts, the modeling situation calls for careful monitoring of key sections of the model, e.g., determining whether a particular component is behaving in a manner intended in the design. Specifically, we may know that some process generator performs well under general operation or even under operations reminiscent of those used in some model. Yet, in sufficiently complex circumstances we cannot tell that the model component is behaving very typically <u>within</u> the given modeling context.

The usual recourse here is to clutter up the dynamics of the model with monitoring code or to otherwise increase the programming burden, e.g., in GPSS, to add extra QUEUE and DEPART pairs (with associated QTABLEs) or in SLAM, with OTPUT routines.

The context, as we are defining it in Figure 3, allows a more sophisticated approach: the process activities can be monitored by a processor which shares random numbers and carries out calculations in parallel. For example, in GGC, we have augmented a number of the standard GPSS blocks so that options not available in GPSS become available. One of these allows easy access to the data input from process generators so that information fed into a block from such sources can be routed onto the net and analysed by what we are calling an "ancillary processor". The code for the model dynamics, thusly, is comparatively uncluttered, and even some increase in efficiency ensues since the model dynamics executor need not attend to some monitoring activities that conventional schemes require.

In AI-oriented modeling, process generators may be created, i.e., generated by the model itself, so that it is impossible to pre-test them, and monitoring them with respect to their usual operation is also impossible (at least until after the fact). The scheme being outlined here obviates some of these unnicities. Hitherto, we have had to rely on relatively difficult processing in Fortran for this kind of work, and, accordingly, we have not pursued to the level desired. A role of S seems apparent.

As promised above, we have some comments on the use of Lisp for numerical computing. In the first case we relate, OPS5 was used for checking random number sequences with respect to their distributional qualities (i.e., goodness of fit tests using ChiSquare). The code is extremely easy to understand, highly parallel in its approach, but so far is quite lethargic in performance on a non-distributed basis.

Accordingly, and for test purposes, we have performed some of the same tasks in both interpretative and compiled mode in Lisp proper. An attempt has been made to program in purely functional style as noted above. The compiled code runs approximately five times faster than the interpretative code, which in turn is faster than OPS5.

In addition to this study with Lisp, we have developed a general linear model package, which can be used in post-run analyses, now, and perhaps with stopping rules in an ASP-based conjunction with our simulation package, GGC. At the time of this writing we have not compiled this code, but in interpretative mode, it performs poorly in comparison to essentially the same code in Fortran. It seems, however, that much of the delay may come from output routines, and we have already taken some steps to improve performance in this category. This Lisp code is easy to modify and a number of "user friendly" elements have been introduced.

An important upshot of this work with Lisp is that it represents a possibility for upgrading the model builder unit so that it can become more self-contained. Some of the behavior hitherto thought to belong only to the unit containing S may be done in the builder unit. For those seeking less AI aid in developing models, the unit containing the S subsystem should also allow entry as well as data analyses, thus extending the system's capabilities for meeting the simulation practice we noted earlier.

The Knowledge Base

Above, we discussed a Lisp-based semantic network integrated with programs that output data that can be processed by Prolog. The Prolog end of this affair is being tied into a version of Prolog which manipulates relational databases. Lisp components related to this have been a focus in two recent developments. In the study by Autrey [29], the focus was on using descriptive information in fuzzy logic for model building and analysis. In the study by Rowe [30], model building capabilities, through incremental refinement have been emphasized. Particularly in the latter study, the role of the KB has been emphasized, in integration with the builder. This approach represents but one avenue toward interaction with the KB.

Another approach centers on PSL/PSA. PSL/PSA represents a tried-and-true environment in which to develop systems, computerized or not [31]. PSL/PSA now runs under a variant of UNIX on the Apollo workstation and Berkeley 4.2bsd UNIX, and we are in the process of porting it to the UNIX5 environment of the 3B system. That such an environment could prove useful for a base for the SE is seen in several ways, perhaps most keenly in some recent papers.

First, in a paper by PSC Inc. [32], the role of simulation modeling in system design is discussed. A stand-alone simulation capability seems more than merely implied in it. This paper also has as target the development of intelligent systems, thus incorporating elements of AI into the perspective. The system is highly graphics oriented, also, and feasibility studies are underway to assess grafting this kind of capability into the SE picture. The work was done in SIMSCRIPT and seems to include a number of features reconcilable with the PSL/PSA approach to documentation.

Second, Bodart [33], again operating in a systems development context, is concerned with evaluation of requirements in a context more tightly bound to PSL/PSA. The phrase, "The Evaluation of Requirements Feasibility by Simulation Generation", which appears early on in his slide presentation, is very revealing toward the ends we are discussing here.

Third, in our own studies, we discussed possibilities for the SE in relation PSL/PSA [25]. In hindsight, we addressed issues such as: integrating of AI techniques into the SE context, with a thorough review of the role of ASP and Prolog in relationship to PSL/PSA-style Knowledge Bases; the connections between our table-processing work in ASP and Prolog and PSL/PSA facilities for manipulating the KB, entering and extracting information from it; and relationship between a detailed design language (useful for writing HELP blocks and event routines in languages such as GPSS and GASP, respectively) and spin-offs from PSL/PSA.

SUMMARY

We have described on-going research on a SE in which we are exploring a distributed system approach initially using a homogeneous AT&T 3B-Net context, which will later be expanded to a heterogeneous system of special purpose processors. The appropriate distribution of various components of the environment among the processors is being explored along with the incorporation of AI tools.

REFERENCES

[1] Henriksen, J. O. "The Integrated Simulation Environment (Simulation Software of the 1990's)." Operations Research, 31, 1053-1072.

[2] Waite, W. Implementing Software for Non-Numerical Applications. Prentice-Hall, Englewood Cliffs, NJ, 1973, 510 pp.

[3] Reilly, K. D., and J. H. Barrett. "The ASP System: The Friendly Front." Proc. 1984 ACM Southeast Regional Conference. Atlanta, GA, April 25-27, 1984, 124-32.

[4] Barrett, J. H. and K. D. Reilly. "Non-Numerical Software Development Studies." Proc. 19th SE ACM Conf., April, 1981.

[5] Barrett, J. H. and K. D. Reilly. "Realization of a Translator for Janus." Proc. 20th SE ACM Conf., April, 1982, 223-225.

[6] Barrett, J. H. and K. D. Reilly. "The Making of ASP: A Language Development Facility." J. Ala. Acad. Sci., 54, 3, 192 (Abstract). Also a presentation at the 60th Ann. Mtg. of the Ala. Acad. of Sci., Tuscaloosa, AL, April, 1983.

[7] Eades, C., Reilly, K. D., Barrett, J. H. and C. Minderhout. "The Barrel Concept: A Study in Language System Development." Proc. 20th SE ACM Conf., April, 1982, 168-171.

[8] Minderhout, C., Reilly, K. D., Barrett, J. H. and J. Gibson. "Use of Barrel in Applications Studies." Proc. 20th SE ACM Conf., April, 1982, 172-175.

[9]   Feigenbaum, E. A. & P. McCorduck. *The Fifth Generation: Artificial Intelligence and Japan's Computer Challenge to the World*, Addison-Wesley, Reading, MA, 1983, 275 pp.

[10] Reilly, K. D. and P. B. Rowe. "Modeling Abnormal Associative Verbal Memory." Conference on Linguistics in the Humanities and Sciences. Birmingham, AL, April, 1984. In: Battistella, E., (Ed.), *Papers on Computational and Cognitive Science*, Indiana U. Linguistics Club, Bloomington, IN, 1984, 87-94.

[11] Reilly, K. D., Salah, A., Rowe, P. B. and P. H. Morgan. "Multiple Representations in a Language-Driven Memory Model." Conference on Linguistics in the Humanities and Sciences. Birmingham, AL, April, 1984. In: Battistella, E., (Ed.), *Papers on Computational and Cognitive Science*, Indiana U. Linguistics Club, Bloomington, IN, 1984, 95-111.

[12] Reilly, K. D., M. R. Freese, P. B. Rowe. "Computer Simulation Modeling of Abnormal Behavior: A Program Approach." *Behav. Sci. - J. Soc. Gen. Sys. Res.*, 29, 3, 1984, 186-211.

[13] Forgy, C. L. *OPS5 User Manual*. Computer Science Dept., Carnegie-Mellon, 1981, 57 pp.

[14] Montalbano, M. *Decision Tables* SRA, Chicago, 1974, 191 pp.

[15] Salah, A., Reilly, K. D., and C. C. Yang. *A Logic Programming Approach to Decision Table Definition and Implementation*. Technical Report, UAB Dept. of Comp. & Info. Sci., Nov. 9, 1984, 18 pp. Also a presentation at the Mid-Southest Chapter of the ACM, Gatlinburg, TN, Nov. 9, 1983.

[16] Reilly, K. D., Dilworth, R. L., Campbell, W. J., Fontana, J. M., and D. Zissermann. "Facilitating Higher-Order Function Programming in Lispkit Lisp." *Proc. 1984 ACM Southeast Regional Conference*. Atlanta, GA, April 25-27, 1984, 71-77.

[17] Hooper, J. W. and K. D. Reilly "An Algorithmic Analysis of Simulation Strategies." *Int'l J. Comp. & Info. Sciences*, 11, 2, 1982, 101-122.

[18] Hooper, J. W. and K. D. Reilly "The GPSS-GASP Combined (GGC) System." *Int'l J. Comp. & Info. Sciences*, 12, 2, 1983, 111-136.

[19] Ziegler, B. P. "Modelling and Simulation Methodology: State of the Art and Promising Directions." In: Dekker, L. et al. (Eds.). *Simulation of Systems '79*. North Holland, N.Y., 1980.

[20] Ziegler, B. P. *Multifacetted Modelling and Discrete Event Simulation*. Academic Press, N.Y., 1984, 372 pp.

[21] Pritsker, A. A. B. and C. Pedgen. *Introduction to Simulation and SLAM*. Halsted, N.Y., 1979, 588 pp.

[22] Spence, D. *Implementation of a Systematic Approach to Monte Carlo on a Minicomputer*. Masters Project Paper, UAB Dept. of Comp. & Info. Sci., May 11, 1979, 239 pp.

[23] Parsons, C. T. *An Investigation into the Further Development of a Joint Simulation System*. Masters Project Paper. UAB Dept. of Comp. & Info. Sci., Feb. 1982, approx. 300 pp.

[24] Garner, C. M. *Structured Systems Analysis Techniques Applied to Combined Continuous-Discrete Simulation*. Masters Project Paper. UAB Dept. of Comp. & Info. Sci., April 13, 1982, approx. 200 pp.

[25] Reilly, K. D., et. al. "Software Development Studies: A Simulation Environment Perspective." *ISETT 1984 Conference*, Ann Arbor, MI, Ref. No. M0657, 16pp.

[26] Becker, R. A. and J. M. Chambers. *An Interactive Environment for Data Analysis and Graphics*. Wadsworth, Belmont, CA, 1984, 550 pp.

[27] Becker, R. A. and J. M. Chambers. "Design of the S System for Data Analysis." *Comm. ACM*, 27, 5, May, 1984, 486-495.

[28] Chambers, J. H., Cleveland, W. S., Kleiner, B. and P. A. Tukey. *Graphical Methods for Data Analysis*. Wadsworth, Belmont, CA, 1983, 395 pp.

[29] Autrey, J. C. *Application of Fuzzy Set Theory to Simulation and Control*. Masters Paper, UAB, 1984, 133 pp.

[30] Rowe, P. B. *Towards a Simulation Environment for GPSS Models by Incremental Model Refinement*. Masters Project Paper. UAB Dept. of Comp. & Info. Sci., 1984, In preparation.

[31] Teichroew, D., and E. A. Hershey III. "PSL/PSA: A Computer-Aided Technique for Structured Documentation and Analysis of Information Processing Systems." *Trans. on Software Engineering*, SE-3, 1, Jan. 1977, 41-48.

[32] PSC, Inc. "Simulation Modeling as an Integral Part of System Specification." *ISETT 1984 Conference*, Ann Arbor, MI, Ref. No. M0637, 26pp.

[33] Bodart, F. and J. M. Leheureux. "Simulation and Prototype Generated from the Requirements Data Base." *ISETT 1984 Conference*, Ann Arbor, MI, (preliminary document), 26pp.

Applied meta-analysis: A procedure for speeding innovation by transferring scientific knowledge more quickly

Douglas Macpherson, Angelo Mirabella
Army Research Institute for the Behavioral and Social Sciences

ABSTRACT

Currently the transfer of scientific knowledge to the applied community can take decades. The system presented here can eliminate the lag. One portion of the system captures the informal rules for performing a complex activity such as simulator design by an expert in an "expert system." The second portion of the system attempts to validate the expert's rules and to discover new rules through "meta-analysis" of the existing research literature. The goal is to provide the practitioners (simulator designers) with access to a computerized expert and to the latest developments in the training literature in a usable form. The system, however, is appropriate for any applied area where there are experts and a research database.

INTRODUCTION

The Army Research Institute is developing a decision support system (DSS) for designing training devices. It will include a rule-based expert sub-system which takes information about training tasks, conditions and standards and then recommends how to design and use devices in programs of instruction. A major tool for developing the DSS, and the subject of this paper, is decision analysis - a combination of meta-analysis and knowledge engineering. Meta-analysis is a set of procedures for translating the findings of a large number of research studies into principles which in turn can be used either to build or verify the rules in a knowledge engineering system. Knowledge engineering, used similarly, is a set of procedures for structuring the knowledge of experts. This paper will describe why these are needed, what their major steps are, how they can be applied to developing training device guidance, and what further R + D are needed.

THE GENERAL PROBLEM

This research is concerned with transferring knowledge from the laboratory to the user. Currently, many specialties, applications personnel (APs) learn applied science, then develop "rules of thumb" for applying their knowledge on the job. But, once trained, they apply new scientific developments slowly. They seldom apply research findings directly because it is difficult to integrate and understand the applied implications of multiple independent studies.

The current scientific knowledge flow is displayed in Figure 1 with the phases of the process shown as column headings across the top of the chart. The process may be described, with considerable simplification, as follows:

o Researchers generate independent studies on the basis of prior research. The research column of Figure 1 shows researchers publishing on Topics N, P ... in their discipline.

o Academicians, integrate the findings of the studies and formulate conclusions. The initial formulations may be disputed and revised. Thus the integration column displays multiple attempts to integrate topics N and P as Formulation J with a feedback loop.

o Educators incorporate the new knowledge into instruction and textbooks when they have confidence in the new formulation. The training column shows Group F being trained in formulation J, a formulation that differs from that taught to the prior Group E. Examples: MDs trained before and after the development of germ theory, electrical engineers trained before and after the development of transistors.

o APs are educated in the formulations derived from the research data base, not necessarily in the research data base itself. As a result the APs may lack the skills required to use the research. Thus Figure 1 does not show a direct path from research to Group F, the group undergoing training at any time, but only an indirect path through the formulation.

o APs then learn the art - the rules of thumb - of their specialty on-the-job. At the same time, they apply their education to develop new rules. However, some of these rules may appear unrelated to the scientific knowledge underlying the applications field. As a result new personnel may have to apply the art by rote because they don't understand its basis. The different thickness of the knowledge flows between Group F and the earlier Group E in Fig. 1 indicates that Group F receives more information from the prior groups than it transmits to them.

o For the remainder of their working lives the APs learn the new science of their discipline from secondary sources e.g. textbooks, trade journals, new people entering the field etc. They then develop new rules on the basis of that information. This link is shown as connector 1.

The implications of the above analysis are:

o The formulations initially learned by applications personnel are based on research that is years out of date by the time they apply it.

o APs tend to lack the education and the time to discover and utilize new scientific developments in their fields.

o APs may tend to retain the art after its use has become inappropriate and apply the art by rote because they fail to understand the basis for the it.

Thus the research integration process (Formulation J in Fig. 1) is a major bottleneck in the transfer of this knowledge to applications personnel. The decisions support system described in this paper, and particularly its use of meta-analysis may break the bottleneck. A second problem involves training personnel in the art of the specialty. The decision support system provides an excellent solution to this problem.

DECISION ANALYSIS

Decision Analysis (DA): The organization and analysis of existing knowledge to generate scientific laws and rules of application. The form of DA developed in this paper combines a rigorous method (knowledge engineering) for determining expert opinion (the art) with an equally rigorous method (meta-analysis) for examining the research literature (the science) to produce a single system for aiding decisions in a technical field.

Knowledge Engineering

As stated above, APs may develop rules/guidelines that appear unrelated to the underlying scientific knowledge. For instance, an expert electronics technician may not reach for his schematics upon reading a complaint. He may, instead, turn on the equipment, quickly sniff as he opens the cabinet, and pass his hand over the circuit board. The sniff can determine if a coil or a capacitor has failed, and the laying of the hands can locate a shorted transistor. Thus the technician may be able to diagnose the problem without using any of the formal knowledge that he has been taught.

Knowledge engineering is a rigorous method for determining the knowledge and rules that experts use to reach decisions in their field. Knowledge engineering then uses computers to mimic human experts by following the same decision rules. The approach has been to "pick the brains" of experts in a somewhat brute force manner. Davis and Rich (1) report that the approach has worked when the area of expertise has been well defined within mature and well codified disciplines such as medicine, chemistry, and geology. The approach may also be productive in less mature disciplines (e.g., training technology) but other approaches may have to be developed to cope with disagreements among experts.

Research in knowledge engineering has demonstrated that the human decision making can be modeled as sets of independent but interlinked rules which the expert can change relatively easily. The research has also demonstrated that a computer system can be created that mimics the decision process of an expert provided that the task is cognitive rather than perceptual, has well defined boundaries, uses professional knowledge rather than "common sense," and can be taught (1).

Practical knowledge engineering experience has demonstrated that an expert in the field must be available and vitally interested in the project. The expert must work with the knowledge engineers and the computer system until satisfied that the system portrays his/her thinking process. The resulting systems frequently mimic approximately 90% of the expert's behavior. They can also explain their decisions by displaying the rules they used to reach the decision, even using the language of the specialist who created the rules. (1) Thus the systems not only mimic the thinking process of the expert, they can teach users to mimic the thinking of the expert. Like human experts these systems are receptive to updates as new knowledge becomes available (i.e., support) in that their rules can be changed relatively easily.

The unsupported knowledge engineering system can have several problems. The expert may be using rules that are out of date or just wrong. More importantly, the unsupported knowledge engineering system lacks one of the major attributes of the expert. It cannot modify its behavior when new knowledge becomes available. As a result it becomes outdated. Thus the expert and the knowledge engineer must meet periodically to update the system and provision must be made for distributing the updates to the users.

Meta-Analysis

Validating the expert's rules and discovering new knowledge requires systematic search and analysis of the research literature. A method for examining the research literature, meta-analysis, was first developed by Glass (2). Meta-analysis is a set of statistical techniques for summarizing multiple independent experiments which are derived from survey research theory. The techniques rest on principle that the experimental effects from different experiments can be transformed onto the same scale. This allows each study conducted in a domain to be treated as an observation in a random experiment which samples the entire domain which, in turn allows the results from different experiments to be combined. Meta-analysis has been criticized, but the criticisms have been answered by improving theory, statistics, and procedures. The procedure has been of sufficient interest that half the issues of the Psychological Bulletin to the extent for 1981-1982 contained meta-analyses or reports related to meta-analysis. Other journals which regularly print meta-analytical research include the American Psychologist, Review of Educational Research, and the American Educational Research Journal. Thus meta-analytical research has been accepted by the most rigorous journals in psychology and education.

To make it easier to carry out a meta-analysis Lawton (4) has developed the Computer Assisted Design System (CARDS). CARDS accepts the results of research studies that use a variety of statistical techniques, transforms them into a common metric, and combines them to produce an estimate of the magnitude of the effect and its variability.

In exploring the research literature with CARDS the user learns what is and is not known about the variables of interest, as well as how important the variables may be. Thus the scientist user can determine what research remains to be done while the practitioner learns the importance of the variables being investigated.

A RESEARCH-DRIVEN DECISION SUPPORT SYSTEM FOR SCIENTIFIC APPLICATIONS

The constraints on knowledge engineering suggest that by itself, it is not feasible for a discipline as primitive as behavioral science. In contrast, decision analysis, as used herein, is viable since it utilizes "hard" experimental data to determine the validity and generality of the rules provided by the experts. The method can even be used to determine what is not known and requires further research. This approach is diagrammed in Figures 3 and 4.

The proposed enhanced knowledge transfer system is presented in Figure 2. This figure shows the earlier groups (E, D) supplying their art (their rules and the reasons for them) to the database management team for a knowledge engineering system. At the same time the scientific literature is supplied to the database management team via an electronic network. The database managers abstract the new information and convert it into a format suitable for rapid scanning by the users and for CARDS (i.e., the output of step d under meta-analysis above) then places it into storage. The stored information is accessed by CARDS operated by the users, including the prior groups, through an intelligent interface. The bidirectional arrow connecting the engine with the database indicates that expert users can also insert rules into the database. This feature allows the system to adapt quickly to the changing environment without expanding the database management function and also provides all users with the new insights of the experts instantly.

The DSS offers a means for training personnel entering the field from relatively unrelated disciplines. The knowledge engineering sub-system displays the rules applicable in each situation and can offer the background information behind each rule. CARDS not only summarizes the literature behind the rules but displays the variation in the studies, thereby depicting the generality of the rule being investigated. Thus new personnel can learn the limits of the art and can employ common sense in their application of the rules.

Our description of scientific and engineering knowledge flow indicated that, to be fully productive, the people concerned with the application of scientific knowledge, such as engineers and medical technologists, require access to the developing art and integrated scientific knowledge of their subdisciplines. The access will be provided with a network of systems which use common rule and scientific data bases to capture the art and science of the subdiscipline. The expectation is that the CARDS rule base will apply meta-analytical procedures to the scientific data base (the output of meta-analysis step d) in order to find and integrate the research findings appropriate to the problem being presented by the user. The intelligent interface to CARDS will then translate the findings into decision rules for entry into the knowledge engineering rule base.

The result of such a system will be immediate diffusion of both the art and the science in any applied field, as shown in the lower portion of Figure 1. This result can yield a savings of decades in the application of some scientific knowledge and will multiply the effectiveness of the best APs by making their techniques available to all, including the trainees. The costs include APs training, initial rule and data base construction, hardware leasing, and data base maintenance. Individual users would pay the terminal/connect costs.

AN APPLICATION FOR A DECISION SUPPORT SYSTEM

Orlansky (5) reported that more than three hundred million dollars are spent each year on the procurement of simulators in the Department of Defense. In the Army, officers and enlisted personnel without relevant training or experience are assigned to training devices procurement. Their support is limited to Army Regulations (ARs) and Army Pamphlets (PAMs), contractor recommendations, and the art existing at their location. These users would benefit greatly from a decision support system that would give them access to the training decision rules used by experts or determined by research to specify the design of training devices.

There are few widely recognized experts in the training domain, and these experts are not available outside their own organizations. The few training experts that do exist are not equally expert on all training subjects. A computerized decision support system combining the knowledge of several training experts and the knowledge embedded in the experimental training literature can make the training knowledge widely and easily available. This scientific applications decision support system will utilize recent advances in knowledge engineering to capture the art of training experts. Advances in artificial intelligence programming and meta-analysis will allow automatic integration and application of the relevant aspects of the research literature to problem solution.

The research program which is required to develop the system has two major components. One component attempts to capture the current art of cost-effective training device design in a system that can be used by the personnel responsible for training device specification development. The second component attempts to relate the research literature to training device design. The second component has the multiple goals of validating the art, discovering new rules, improving theories, and specifying critical experiments to resolve training device design issues. Thus the second component allows the DSS to adapt intelligently to a changing environment.

The Art

The development of the art portion of the DSS requires a computer, a knowledge engineering computer program for organizing and using the expert's rules, and access to experts in the training domain.

As shown in Figure 3 the procedure to develop this component requires:

a) reading secondary sources such as ARs, PAMs, and existing analytic models for guidance concerning the specification of training devices;

b) reducing the guidance to rules acceptable to the Knowledge Engineering System (KES);

c) preparing other rules from knowledge of the art and science;

d) installing the rules in the rule base, then testing the system to determine its weaknesses;

e) using Delphi Groups and individual interviews to acquire information for improved rules;

f) repeating steps b-e until a satisfactory system is developed.

This component is under development. Singer (6) is producing a knowledge engineering system using a rule base and any mathematical models that are developed. However the system operates on a VAX computer which is not portable and cannot be field tested easily because of problems in acquiring field computer communications support. An alternative is to use one of several microcomputer knowledge engineering systems which do not require the input of rules, but induces them from examples, then transforms the results into a form which can be implemented widely. Such systems will allow the VAX/KES rule base to be transported to the field, then used and improved by experts, even those who cannot formulate the rules in knowledge engineering terms. Thus these systems will increase the size of the pool of usable experts from those who are verbal and have access to data lines to the entire population. In addition they provides a relatively inexpensive method for field testing the system.

We note that the art component of the DSS can be generalized to cover many action officer, as well as highly skilled enlisted and warrant officer positions. In principle, the procedures described above can be applied to any well defined problem area where recognized experts produce similar decisions. We intend to demonstrate this capability by expanding the art portion from training device specification to a training system design expert. Our approach is to create modules which provides expertise in related technical areas in the training domain, thus gradually building a system with more expertise then any one expert. Eventually, we will attempt to combine

modules and thereby allowing search of the larger decision space for more effective trade off.

The Science

The ultimate objective of this component is to develop a methodology for transferring scientific advances rapidly and effectively to the user. It rests on the premises that the science behind any technology develops continuously, that the relevant research cannot be located for all problems in advance of the applied requirement, and, as stated previously, the applied personnel may lack the ability to integrate and understand the implications of the research for the problem. The immediate objective is to develop AI controlled meta-analysis procedures that will assist in integrating areas of the research literature and in development of a theory of simulator design. Thus the meta-analysis procedures will perform the research integration and help the user to understand the implications.

The purpose of meta-analysis is to find consistencies in dependent variables responses to similar situations in different experiments. Thus the concept is analogous to the convergent validity construct. It differs in that it is retrospective, summarizing the results of existing experiments rather than being an experimental design technique. As shown in Figure 4 the proposed developmental effort will utilize meta-analysis to test the validity of the "rules of thumb" employed to specify training devices. Meta-analysis will also be used to explore the training devices research literature in an attempt to develop valid new rules of thumb. Finally, the figure indicates that the CARDS system will be designed for use by training devices personnel in analysing the research literature and developing rules for application to their specific problems. Thus the training devices personnel will be able to apply the information from new research as soon as it is entered into the database. By implication CARDS will also have the capability to aid in developing theory and to suggest directions for research thrusts. It will achieve the former by demonstrating regularities in results across experiments in terms of the size of the independent variable effects. It will accomplish the second by mapping the training devices literature in terms of overlapping studies and locating combinations of variables that have not been utilized in research but that are required to connect and generalize existing experiments, thereby allowing critical tests of hypotheses concerning relationships among variables.

System development will require:

a) locating and abstracting research literature relevant to training devices;

b) preparing meta-analysis abstract;

c) translating abstracts into a data base format that tolerates missing elements;

d) performing meta-analyses that will determine the scientific support for the rules in the "art" component;

e) examining the research data base via meta-analysis to generate new rules, then suppling the rules to the art component for insertion at step d in that development process;

f) interfacing the two components so rule generation and testing procedures can be automated;

g) preparing a human-system interface that will allow training device personnel to query the meta-analytic data base and obtain usable responses/rules.

The first step above requires access to the training research literature. This is a highly diffuse body of information which is not available in any one location. Appropriate articles are usually located by searching a variety of computer databases and by examining the references provided in the applicable research. Neither method provides the information required to perform meta-analyses, nor even to assure identification of appropriate studies.

Hays and Singer have prepared a computer accessable version of Ayres and Heinicke's Annotated Bibliography of Abstracts on the use of simulators in technical training (7). These abstracts present the

experimental research in the field of training
simulators in a consistant 10 section format that
provides a full description of the devices, the
situation, and the independent and dependent variables, as well as a partial presentation of the key
results. Thus these abstracts provide the information
required to locate experiments and some of the information required to perform meta-analyses.

RESEARCH ISSUES

The proposed DSS represents a melding of computer
science, cognitive science, training research, statistics, and psychometrics. Advances in each will be
required before the DSS can be used by novices to
devise new training device specifications from new
research. However the advances are not required to
develop a less responsive, capable and intelligent
system which requires a more capable staff.

Current Research and Development Requirements

Knowledge engineering. An experimental form of
KES is under development. However the rules are being
derived from the Army Regulations and training texts.
They rely more heavily on logic to fill in the gaps
in knowledge than on personal expert knowledge of
successful and unsuccessful training simulators. Thus
the system will represent one operationalization of
the model implicit in the training literature. This
approach is expected to produce a useful system for
training device specification in 1985 even though
many rules may not have demonstrable empirical validity.

A problem with system development has been that
knowledge engineering systems require that only one
expert be used so as to provide a consistent rule
base, yet no expert is known who has the requisite
breadth and depth of knowledge to produce or even
evaluate a training devices specification KES. Thus
Singer (6) has proposed development of a new knowledge engineering procedure which would utilize Delphi
techniques to produce and synthesize the opinions of
multiple experts into a consistant set of improved
rules. This effort will require the development of
Delphi protocols, the identification of appropriate
experts, their assembly into Delphi groups and production of new rules. This effort could be completed
in 1985 and would produce an expanded KES system
which has been validated in terms of expert opinion.

Meta-analysis. The applied research issue of
primary concern involves the validation of the KES
rule base. In this effort the abstract data base will
be extended to include new research and all data from
the cited research. The abstract data base will then
be input into CARDS. Finally the KES rules will be
transformed into hypotheses suitable for CARDS and
meta-analysis will be attempted. It is expected that
many rules will prove to be untestable because appropriate experiments are lacking. The output of the
effort will be guidance for the design of research
plans and the scaling of each scalable rule for magnitude of effect. The latter information will be used
to alter the KES control structure and annotate the
KES rules.

An effective computerized meta-analysis system
should provide the capability to enter new studies,
search for research of interest, examine it, and
enter it into the meta-analysis data base without
difficulty. The current approach requires that new
studies be entered using special codes to identify
items to be indexed, then generating a second, read-
able copy. The following steps are required to perform a meta-analysis:

o look for words and phrases of interest in the
index;

o direct the computer to the pages of interest;

o read the abstract containing the word of interest;

o manually copy the data of interest;

o switch from the reading file to CARDS;

o manually enter the data into the CARDS data
base;

o and finally request the meta-analysis calculations.

The approach is labor intensive, does not permit
complex searches, and permits errors in preparing the
indices and in transferring the data to CARDS. A
software development program to integrate the
abstracting effort with CARDS has been started. The
integrated system will generate indices and
subindexes by topic and update the CARDS database. It
will also allow the user to employ CARDS to perform
searches against complex criteria, read them, and
enter them into the analysis without leaving the
CARDS environment.

We propose to generalize meta-analysis from a
procedure limited to determining which individual
independent variables have significant effects into a
procedure for "mapping" research domains using experimental design principles from psychophysics. In
principle, the procedure requires mean effect size
determination of all the independent variables of
interest. Next the research literature is re-examined
to find overlapping experiments, experiments that use
the same level for some experimental condition such
as the control condition. The effect sizes for the
significant experimental manipulations in each experiment are then treated as scaled values and the
scales derived for the various experiments are mapped
onto a common scale and checked for interactions. The
mapping operation will provide the information required to construct a model predicting the effects of
the independent variables in the domain. The model,
in turn, will allow development of a theory of training simulator specification. Even if the research is
insufficient to develop a model the meta-analysis
effort will identify the experiments and the work
program required to develop, validate and complete
the theory.

A severe limitation in traditional meta-analysis
has been the requirement for hundreds of experiments
in the field of interest. The requirement exists
because the techniques use normal probability statistics to perform analyses incorporating large numbers
of moderator variables. In Efron's Bootstrap (9), the
error of estimation for the population parameters is
calculated by sampling repeatedly from a pseudo-popu-
lation which was created by replicating the existing
sample a large number of times. Efron (10) states
that this non-parametric method appears to allow
greater, precision in the estimation of population
parameters than normal probability theory, thereby
reducing the sample size required to achieve significance. As a result it may provide useful analyses
with the numbers of experiments available in the
training devices literature. Since the behavior of
the Bootstrap as a function of the number of observations and in the presence of sampling and measurement
errors is not known, a series of simulation experiments will be conducted to compare the performance of
the Bootstrap with normal probability distribution
estimates. The immediate results of the effort should
include papers in statistics, testing, and
meta-analysis documenting the relative performance of
the procedure. If the research is successful than the
Bootstrap procedures will be included in CARDS.

Future Research and Development Requirements

Knowledge Engineering. Frequently, by the time
people have become experts in a field they have coded
and condensed their knowledge to the point that they
can no longer describe how they reach their decisions, much like the bike rider who cannot describe
how he keeps his balance. Thus extracting knowledge
engineering rules from an expert is frequently a
difficult and time consuming effort because the expert has not coded his procedures in a verbalized
rule format. In contrast, Michie (8) claims that
experts can readily identify appropriate solutions to
specific problem situations, and has constructed a
computer system that induces the rules from examples
provided by the experts. However no research has been
reported that contrasts the two approaches. If
pretesting the rule oriented Delphi procedure suggests that training experts cannot verbalize their
simulator design rules, then a situation-oriented
Delphi procedure will be developed and an experiment
conducted to determine the relative merits of the two
approaches in terms of experimenter effort and expert
time. In addition to the effort described in the
prior paragraph, the experiment will require acquisition of a computer program for acquiring and translating situation information into decision rules and
the development of situation oriented Delphi
protocols. The products of this research will be a
research report contrasting the two approaches to

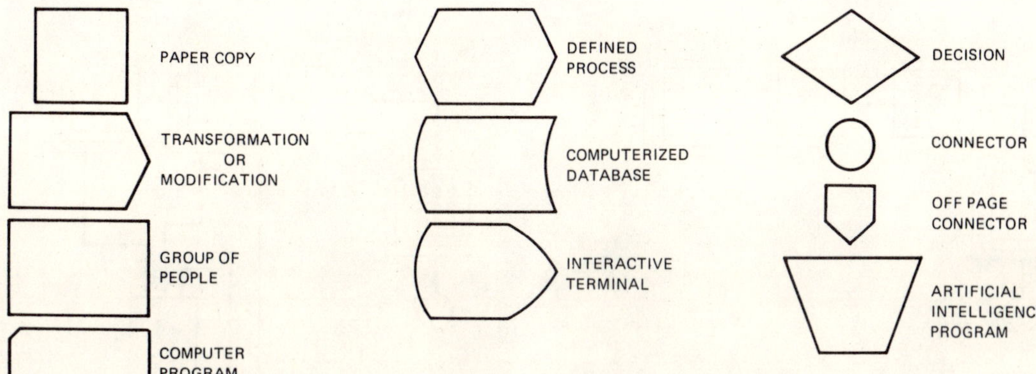

Figure 5. Symbols defined

knowledge engineering, the extended KES system, a self contained microcomputer based KES system which explains its decisions in terms of the situations that produce that decision, as well as the rule based KES system.

Meta-analysis. An issue in the second component is the development of an intelligent interface to assist the user in formulating meta-analysis tests of proposed rules and to answer specific applied problems. These issues in the second component are not crucial to the demonstration system because the users of the demonstration system will have the requisite skills. However the intended users will require such assistance to extend the system.

The final major issue in the proposed research is relevant only to maintenance of the system. The problem is that the training device specification environment changes, with the result that the rules must be changed. The remaining major problem in knowledge engineering system design involves "revalidation" of the system when the rule base is changed because the AI researchers have yet to find an algorithm for checking all possible outcomes that is not subject to a combinatorical explosion as the number of rules increases. The researchesthat will store the tests used to validate the system. Revalidation will consist of rerunning the original validation tests and not attempting to validate for all possible combinations of rules.

SUMMARY AND CONCLUSIONS

An analysis of the flow of scientific information from the bench scientist to the user suggested that there are two major choke points in the transfer of scientific information - the research integration phase and the actual transfer of the integrated information to the practitioner/user. These choke points can delay the application of scientific knowledge by decades. However, recent developments in the field of research integration (meta-analysis) and in computer science (expert systems and networking) suggest that a system can be developed which will provide the practitioner with most of the decision making capability of an expert in the field. Furthermore, the system would be able to integrate new developments in the field in such a way that new theories, decision rules, and research recommendations will be generated.

Consideration of the sums spent on training devices and of the preparation of the personnel assigned to select the training devices suggested that a demonstration decision support system designed to assist in the specification of training simulators could provide a useful tool. The research and development required to produce such a tool are described.

REFERENCES

1. Davis, R. & Rich, C. A tutorial on expert systems: Part 1 - Fundamentals. Melno Park, CA: American Association for Artificial Intelligence, 1982.

2. Glass, G. Primary, secondary, and meta-analysis of research. Educational Researcher, 1976, 5, 3-8.

3. Hunter, J. E., Schmidt, F. L., Jackson, G. B. Meta-analysis: Cumulating research findings across studies. Beverly Hills CA: Sage Publications, 1982.

4. Lawton, G. W. Description of the preliminary development of the Computer Assisted Research Design System (CARDS). in Conference on Artificial Intelligence. Rochester MI: Center for Robotics and Advanced Automation, Oakland U., April, 1983.

5. Orlansky, J. Cost-effectiveness of military training: 6-22-82-6. Arlington, VA: Institute for Defense Analyses, 30 June 1982.

6. Singer, M. An expert system based training device design decision support system, Version 1.0. ARI Working paper 84-9

7. Ayres, A., Heinike, M. An annotated bibliography of abstracts on the use of simulators for technical training. ARI Technical Report Alexandria Va.: U S Army Research Institute for the Behavioral and Social Sciences, (in press).

8. Michie, D. Personal Communication, April, 1983.

9. Efron, B. Bootstrap methods: Another look at the jackknife. Annals of Statistics, 1979, 7, 1-26.

10. Efron, B. Nonparametric estimates of standard error: The jackknife, the bootstrap, and other methods. Biometrika, 1981, 68, 3, 589-599.

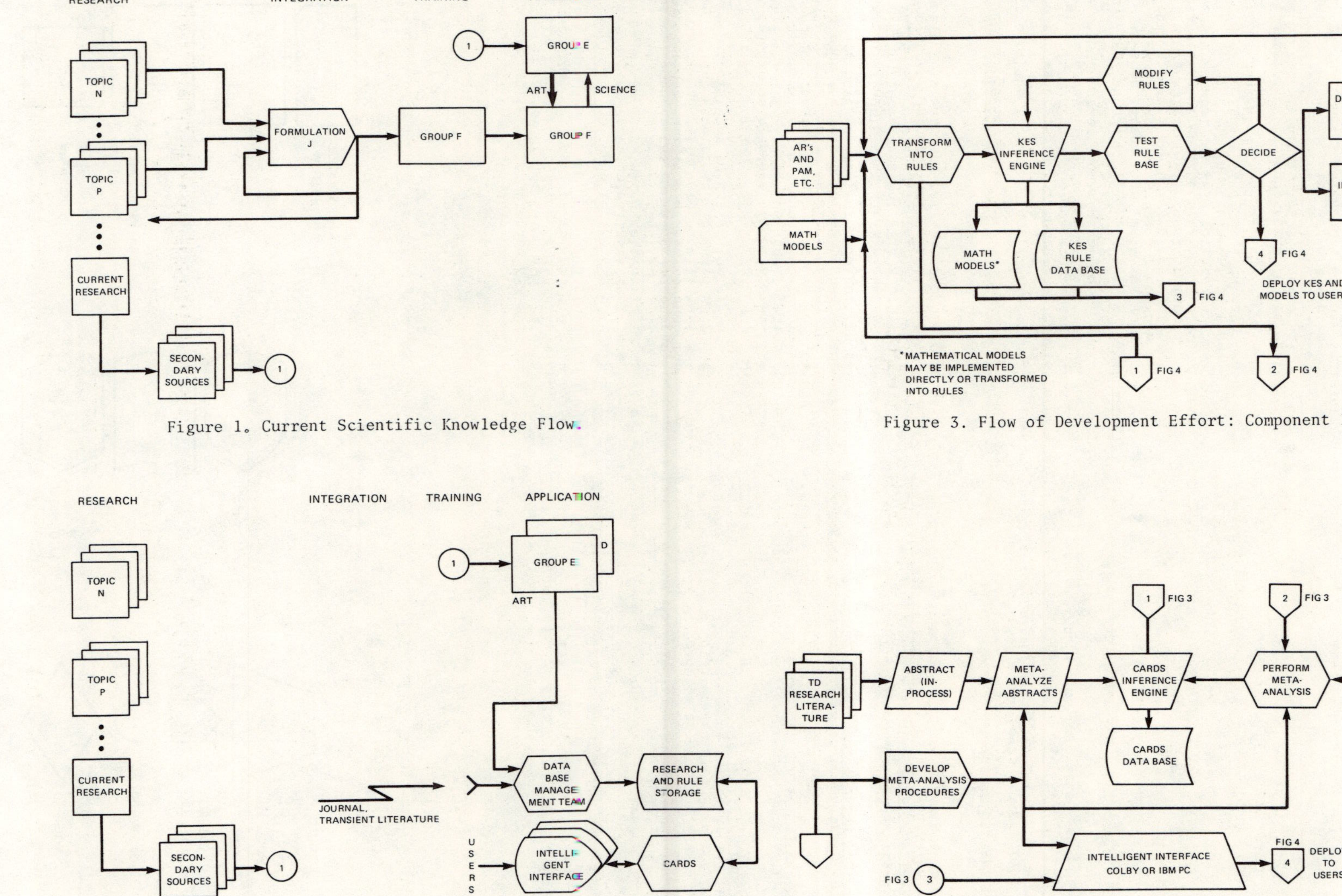

Figure 1. Current Scientific Knowledge Flow.

Figure 2. Proposed Scientific Knowledge Flow.

Figure 3. Flow of Development Effort: Component 1.

Figure 4. Flow of Development Effort: Component 2.

Syntax programming, expert systems, and real-time fault diagnosis

W. ROBERT COLLINS* and STEFAN FEYOCK*

Department of Computer Science
College of William and Mary
Williamsburg, Virginia 23185

Abstract

Of great interest today in real-time programming is the **pilot's associate**, a computer system whose purpose is to mitigate the complexity of a pilot's job. The research we report concerns the rapid development of fault diagnosis expert systems to be integrated into the general simulation models at NASA Langley Research Center. The approach we use, called **syntax programming**, addresses the efficiency and interface problems found in systems produced in standard Artificial Intelligence languages as LISP and PROLOG and executed on standard simulation hardware. Syntax programming is a style of programming that incorporates ambiguous attribute grammars, parser generator systems, and pre-built program skeletons to produce detailed HOL programs (Pascal, in this case) whose behaviors are specified by BNF-like grammars.

Syntax Programming

Syntax Programming (or SP) may be defined as the use of grammatical specifications, a parser generator system, and a HOL "library" to construct programs called SP programs. SP was coined by Feyock[1] because of the similarities between syntax programming using the tools above and logic programming using PROLOG. Indeed, many of the initial Artificial Intelligence applications of SP were derived from similar capabilities of PROLOG. Research has convinced us that SP cannot replace logic programming, and consequently some SP constructs created to mimic logic programming are no longer in use. There are a variety of applications where SP is as powerful as traditional AI techniques, and many more traditional applications where SP is a surprisingly efficacious technique.

---

*Work funded by NASA grant NAG-1-469, NASA Langley Research Center.

The grammatical specification for an SP program is similar to the rule base of a production system. A grammatical rule of the form

$$\langle lhs \rangle ::= \langle rhs_1 \rangle \ \langle rhs_2 \rangle \ ... \ \langle rhs_n \rangle$$

corresponds superficially to the production

$$\text{IF } \langle rhs_1 \rangle \text{ AND } \langle rhs_2 \rangle \text{ AND ... AND } \langle rhs_n \rangle$$
$$\text{THEN } \langle lhs \rangle$$

An alternate interpretation of the grammar rule might be

$$\text{DO } \langle rhs_1 \rangle$$
$$\text{DO } \langle rhs_2 \rangle$$
$$...$$
$$\text{DO } \langle rhs_{k-1} \rangle$$
$$\text{IF } \langle rhs_k \rangle \text{ AND ... AND } \langle rhs_n \rangle$$
$$\text{THEN } \langle lhs \rangle$$

How the grammar rule is interpreted is the programmer's decision.

There is an important distinction to be made between these grammatical rules and standard AI productions. Each $\langle rhs_i \rangle$ is itself the left hand side of some other rule. To say that $\langle rhs_i \rangle$ is true means that there is some rule

$$\langle rhs_i \rangle ::= \langle a_1 \rangle \ \langle a_2 \rangle \ ... \ \langle a_j \rangle$$

that has already fired. But the latter rule cannot fire unless there are rules with each of the $\langle a_t \rangle$'s on the left hand side that have fired, and so on. One gets out of this infinite regress in normal parsing by eventually having terminal symbols (as opposed to grammar nonterminal symbols enclosed in $\langle ... \rangle$) in the rules. In SP grammar specifications there are usually no terminal symbols. The process terminates when there are **empty** rules, that is, rules of the form

$$\langle lhs \rangle ::=$$

whose triggering is controlled by their disambiguation semantics.

The semantic component of the specification, HOL code associated with each rule, has two flavors: disambiguation and action. Disambiguation code is used to control the triggering of rules; predicates are evaluated at run time and rules with true values are triggered. The rules are ordered and the first triggered rule is fired. Disambiguation code is indicated by a slash "/" in the first column (whereas rules are distinguished by a left angled bracket "<"). For example,

```
<rotor_symptom> ::= <metal_fatigue>
/ (* fire if tests A and B are both positive *)
/ return := (test_run(A) = positive) and
            (test_run(B) = positive);
```

If this rule is in the subset of rules the parser is able to fire in its current state, then the rule is triggered when the value of the Pascal variable return is true. This value is computed at run time, and corresponds to part of the **context** of normal production systems.

Action code for a rule is executed just before that rule fires. A blank " " in the first column distinguishes action code from disambiguation code or rules. This code may be used for computing actions not amenable to rule-based processes such as I/O, sensor information, or time considerations. From the example above,

```
DO <rhs1>
```

means execute the action code associated with the firing of some rule with <rhs$_1$> as its left hand side. One way to ensure this is by an empty rule:

```
<rhs1> ::=
/ return := true;
  begin
      (* do this action code *)
  end;
```

We may omit disambiguation code whose return is constantly true. Action code may also appear in rules with non-empty right hand sides.

A scenario for building an SP program starts with the construction of a grammatical specification for the application. The general flow of control for the application is dictated by the structure of the application and its possibly changing data base or context. In the grammatical specification, these two items are reflected in the syntax (grammar rules) and the disambiguation semantics (HOL code that controls triggering and firing). During the application process, the context is updated and actions taken. These are reflected quite naturally in the action semantics (HOL code executed concomitant with rule firings).

The grammatical specification together with the appropriate HOL library entries (or even a preliminary version of the expert system program) form the input to the parser generator. We use the **Mystro**[2] system which encompasses a full LALR(1) parser generator, report and file merging capabilities, utility programs for manipulating grammars and attribute references, a parse tree generator, a parse tree transformer, automated code generator generator programs, and libraries of HOL code fragments and skeletal programs. Mystro is extremely user friendly and runs interactively. The parser generator's output is an entire, compilable HOL program that behaves according to the grammatical specification. HOLs in active use are Pascal and LISP. HOLs under research consideration are Ada and PROLOG.

There are distinct advantages to using SP instead of standard application program languages or Artificial Intelligence languages. A general advantage SP has over standard languages is that SP programming is syntax directed. Since one encodes each rule separately, the programming task is automatically broken up in many small programs. It is difficult to see the "large" picture, however, while encoding each rule. In this case, the highly structured nature of the grammatical specification itself, one of great simplicity and clarity, facilitates programming understanding.

The resulting SP programs are efficient. The closest analogy to standard PROLOG- or LISP-based production systems is the following: the firing of an SP rule automatically limits the choice of succeeding rules that can be triggered to those found in the next state. The parser built in every SP program is a very efficient manager of the classes of rules that apply in different states. LALR(1) parsers operate in linear time compared to the exponential time for many standard tree searches. Moreover, the parser is driven by highly compressed parse tables, and is thus space-efficient as well.

Lastly, parsers are capable of introspective behavior. Parsers, and by extension SP programs in general, have their control behavior determined entirely by the parse tables that drive them. These tables are available at run time, and thus provide, at run time, a machine-readable program specification. The full range of questions that can be addressed on the basis of the parse tables driving the program is large[3]. Nonetheless, this introspective capability is not as easy to manage as the rule base for a LISP or PROLOG production system. Flexibility is sacrificed for efficiency. The appropriateness of the compromise depends on the application. In real-time aircraft fault diagnosis efficiency is important.

Aircraft Fault Diagnosis Research

The **Intelligent Cockpit Aids** project at NASA Langley Research Center houses a variety of related projects, one being Artificial Intelligence research in fault diagnosis[4] conducted by the Flight Management Branch of the Flight Control Systems Division. Part of the research in this area involves the use of a simulation model of a physical system to generate the knowledge base for fault diagnosis of that system. Failures are simulated in the model, the symptoms detected from the simulation, and both are stored in a "fault dictionary." The model for the fault diagnosis research system is contained in Figure 1 below.

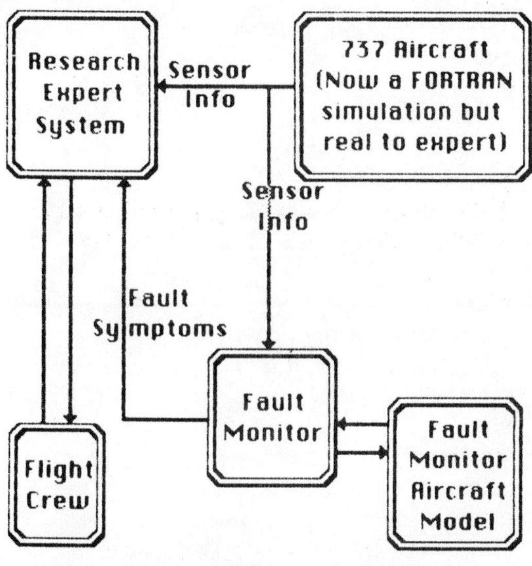

Figure 1

This system consists of an aircraft, a fault monitor, and a research expert system. The simulated aircraft is a Boeing 737, though other models are envisioned. The fault monitor is an existing software package that receives sensor information from the aircraft and compares this information with its own internal model of the aircraft. If the fault monitor finds discrepancies between its expectations and the "real world," it outputs a description of these discrepancies to the expert system fault diagnoser. The expert system will receive inputs both from the sensors and from the fault monitor. It will attempt to produce a diagnosis of the most likely fault on the basis of the sensor inputs, the symptoms from the fault monitor, and, if necessary, feedback from the flight crew. It will transmit its conclusions to the flight crew, and should be capable of further interaction with the crew, for example to explain the basis of its conclusions.

Two varieties of "expert" action have been constructed for this application using SP. The first variety is a procedural, perturb-the-system diagnostic method and the second a complex decision table diagnostic tool invoking past behavior as a predictor of current behavior. Of course, these two systems are only small parts of the overall fault diagnosis expert. Both diagnostic systems are essentially deterministic and involve little reasoning by the expert system programs. We will expand on each.

The Procedural Expert

The procedural method is derived from a transition diagram detailing the process appropriate for an electrical fire. Time is an important factor in determining what is causing the fire. The solution to the problem is a direct consequence of the cause of the fire: land immediately if the problem is intractable or continue to fly in a certain configuration. The expert system was constructed automatically with a translator program **TD Converter** that converts the transition diagram into a grammar and allows usage of a stylized natural language to build the action and disambiguation HOL semantics. Action semantics are essential to the time delays, where pilot actions were contingent on smoke increasing, decreasing, or remaining constant after a "reasonable" wait.

Figure 2 shows part of the transition diagram and Figure 3 shows the grammar fragment constructed automatically by TD Converter.

Nodes in the transition diagram are essentially state markers. All information about a node is contained in the arrows leaving the node. There are two kinds of information on a node: the disambiguation predicate before the perpendicular line and actions to be taken after the line. The disambiguation predicate is used to decide which arrow to take. One predicate will evaluate to true; the actions following it will be done and its arrow taken to the next node. In Figure 2, the starting node is labelled **Circuit Breaker**. If no circuit breaker has tripped, then the flight crew will be instructed to turn off the indicated switches and a computer-controlled reasonable delay will occur. After the delay is accomplished, the status of **smoke** will determine both the actions to be done and the next node.

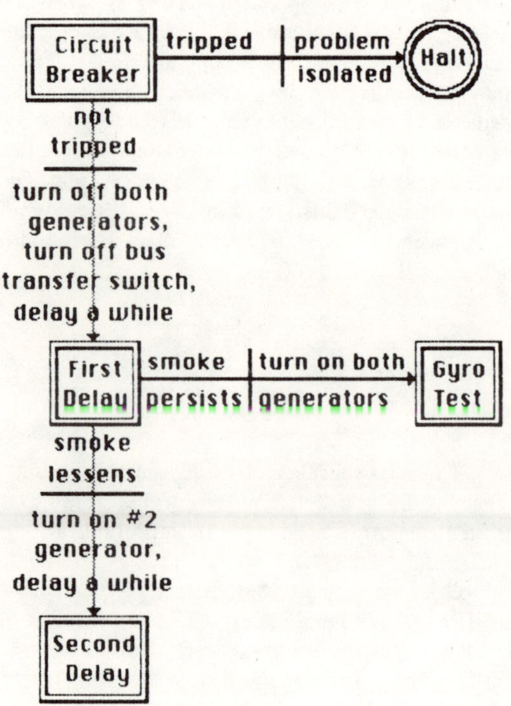

Figure 2

Figure 3 shows the output from the translator program (slightly edited to fit within the margins and for clarity). The rules are written in BNF style, the disambiguating code is preceded by "/" (slash), and the remaining code is action code. The rules are listed in top-down fashion, corresponding to the transition diagram. However, the rules are evaluated bottom up. For example, the second rule

<breaker> ::= <br_qu> <br_no> <1st_delay>

can be translated as

DO <br_qu>
IF <br_no> AND <1st_delay>
THEN <breaker>

that is, **<breaker>** is fired when **<br_qu>**, the breaker query, is done and **<br_no>** and **<1st_delay>** are fired. "Doing" **<br_qu>** means executing the action code associated the rule

<br_qu> ::=

Firing **<br_no>** means that the disambiguation code for the rule

<br_no> ::=

returns true, which it does when the response to the question asked in the action code for the rule

<br_qu> ::=

is N (for no). Firing **<1st_delay>** occurs only in the rules where **<1st_delay>** is on the left hand side, as in the sixth and seventh rules.

```
<breaker> ::- <br_qu> <br_yes>
    writeln('Manual claim: Your problem is isolated.');
<breaker> ::= <br_qu> <br_no> <1st_delay>
<br_qu> ::=
    begin
        writeln('Has circuit breaker tripped?');
        writeln('Answer Y or N.');
        readln(response)
    end;
<br_yes> ::=
/   return := response in ['y', 'Y'];
<br_no> ::=
/   return := response in ['n', 'N'];
    begin
        writeln('Turn OFF both generator switches.');
        writeln('Turn OFF bus transfer switch.');
        delay(a_reasonable_time)
    end;
<1st_delay> ::= <1st_qu> <1_yes> <2nd_delay>
<1st_delay> ::= <1st_qu> <1_no> <gyro_test>
<1st_qu> ::=
    begin
        writeln('Has the smoke lessened?');
        writeln('Answer Y or N.');
        readln(response)
    end;
<1_yes> ::=
/   return := response in ['y', 'Y'];
    begin
        writeln('Turn ON generator #1 switch.');
        delay(a_reasonable_time)
    end;
<1_no> ::=
/   return := response in ['n', 'N'];
    writeln('Turn ON both generator switches.');
```

Figure 3

An alternate view of the second rule is that it gives a recipe: a way to leave the breaker node is to ask the breaker query, receive the answer "no," and then go to the 1st delay node. This is how SP programmers will interpret the grammatical specification.

The preliminary electrical fire transition diagram contained approximately 15 nodes. It took about one hour to prepare a skeletal program containing delay and verify procedures, about thirty minutes to create a grammatical specification from the diagram using TD Converter, about two minutes to execute the Mystro parser generator, creating the expert program, and about five minutes to compile, link, load, and execute that program on sample data. All these times are wall clock times on a multi-user supermini computer. Both TD Converter and the Mystro parser generator are sufficiently friendly to be run with little training. TD Converter contains a stylized natural language translator that creates Pascal code (either menus or normal statements) from user prompts.

The Decision Table Expert

The decision table involved the diagnosis of the cause of an engine malfunction. The malfunction was made manifest through anomalous gauge behavior or symptoms such as vibration and moisture. The decision table was complex for three reasons: multiple values for the same gauge could indicate the same fault; the state of all guages could indicate the potential presence of two different faults; and many potential gauge state configurations were not present in the decision table. (This particular application was initially suggested by an already existing solution in PROLOG[5].)

The gauge state decision table was also incomplete. Prior faults could influence the current fault and multiple fault behavior was not explicated. Neither causality nor past histories of gauge states were amenable to grammatical specifications; these had to be encoded in the action and disambiguation HOL code. As in the transition diagram example, once we realized how to convert the decision table into an SP specification, the process seemed automated enough to program. The program **DT Converter** prompts a user to specify the rows and columns of a decision table, the potential values for each column, and then the actual values found in the table. Given this information, it is straightforward for DT Converter to create a BNF grammar with Pascal disambiguating and action code.

Grammars are reasonably modular, as long as there are no symbol name conflicts. To solve the engine malfunction problem, DT Converter was used to create an SP specification and then that grammar was inserted into a larger grammar that handled

both the simulated history of the engine and the simulated existence of faults. Figure 4 shows a small section of the DT Convertor-generated SP specification.

```
<row>   ::= <vib_y> <egt_f> <epr_f> <n1_f> <n2_f>
            <ff_n_f>
    writeln('Row value is foreign object.');
<vib_y> ::=
/   return := letter_vib = 'Y';
<vib_n> ::=
/   return := letter_vib = 'N';
    •
    •
    •
<ff_n_f> ::= <ff_n>
<ff_n_f> ::= <ff_f>
<ff_n>   ::=
/   return := letter_ff = 'N';
<ff_f>   ::=
/   return := letter_ff = 'F';
    •
    •
    •
```

Figure 4

The long rule corresponds to the production
IF vibration(yes)
    AND egt(fluctuating)
    AND epr(fluctuating)
    AND n1(fluctuating)
    AND n2(fluctuating)
    AND fuel flow(normal or fluctuating)
THEN problem is isolated

The actual problem encountered is not present in the grammar, but is found in the action semantics that inform the user of the problem. Each of the nonterminals in the right hand side of the rule are fired by disambiguating semantics that flow from knowledge of the symptom. In the fragment below

```
<vib_y> ::-
/   return := letter_vib = 'Y';
```

the apparent opacity is due to the fact that DT Converter created this code, and used letters to stand for the values of vibration (present or not). This allows the program to avoid duplication of enumeration literals. A more modern language like Ada would allow clearer code to be generated automatically.

This application pointed out clearly the differences between SP and logic programming with PROLOG. An important part of the problem is the fact that the past influenced the present. For example, if there was a foreign body in the engine in the past, there may be compressor stall now. One can use this information to help diagnose the reason for anomalous gauge behavior. In PROLOG, temporal information had to be carried essentially as a side effect in an extra parameter in all of the functors that defined diagnoses. This led to an unnatural PROLOG program, hard to construct and read. There is no facility whatever in grammatical specifications and parsers for remembering the past state of symbols in the grammar. However, the action and disambiguation code, written in a language amenable to the data structures needed to remember past states, covered the situation handily. The principle here is that when something is hard to code with grammars or LISP or PROLOG (and all languages are awkward at some things), then the procedural HOL available with SP can be used (in the action and disambiguation code). In this sense, an SP program can combine the best of two disparate categories of computer language.

A second major difference between the SP model and the PROLOG model was that in PROLOG, data were stored as PROLOG "facts" in files. The data could have been received interactively, but it would be difficult in a normal programming environment to receive data from sensors or other independently running programs. PROLOG (and LISP) do not generally interact well with outside systems without extensive work. The SP program, written in a procedural HOL, easily interfaces with other programs. In a sense, the parser generator is acting as an SP compiler. It receives a grammar specification with action and disambiguation HOL code as well as the appropriate skeletal HOL program. It produces as output a complete HOL program which can be compiled itself and executed. The low level connections needed to interface with an embedded simulation system can be built into the skeletal HOL program. The SP programmer need do that once, and thereafter ignore that part of the application. This frees the SP programmer to concentrate on the grammatical specification which resembles the problem at the highest level.

References

[1]Stefan Feyock, "Syntax Programming," *American Association for Artificial Intelligence National Conference*, Austin, Texas (August 1984).

[2]Robert E. Noonan and W. Robert Collins, *The PARGEN Users' Guide*, Version 7.0, William and Mary Technical Report 8401, (January 1984).

[3]Stefan Feyock, "Transition Diagram-based CAI/HELP Systems," *International Journal of Man-Machine Studies*, volume 9 (1977), pages 399-413.

[4]Kathy Abbott, "AI Research at NASA Langley Research Center -- Intelligent Flight Management," *AI Magazine*, volume 5, number 3 (Fall 1984), pages 79-80.

[5]Kathy Abbott, "A Fault Diagnosis System for a Turbofan Aircraft Engine," internal report, Rutgers University (Spring 1984).

RBMS - An expert system for modeling NASA flight control room usage

Christopher A. Marsh
Ford Aerospace & Communications Corporation
Artificial Intelligence Laboratory
Houston, Texas 77258

ABSTRACT

With the number of Space Shuttle flights increasing to 24 flights per year, a model to simulate the usage of the Flight Control Rooms (FCR) was required to estimate the resource usage and capacity of the current and new configurations of the Mission Control Center (MCC) at the Johnson Space Center (JSC) in Houston. The goals of this project were to build a model that was fast and easy to run, had outputs that were easy to interpret, and could easily be changed to reflect different situations. Artificial Intelligence (AI) techniques were used in building the model because it was the best way to meet these goals.

This Rule Based Modeling System (RBMS) is a deterministic model written in LISP using the OPS5 inference engine. This paper describes how RBMS works and how it is being used as a tool for estimation of FCR usage and operational capacity of the MCC as a function of the flight manifest and the characteristics of the activities required for each flight.

INTRODUCTION

Projecting the required resource levels has always been a problem of great concern to manufacturing and construction companies. The success or failure of many companies rests with the efficiency of management in dealing with scheduling and resource allocation. It has become evident to NASA that the Shuttle Transportation System (STS) operation is facing this problem. It was clear that some means had to be developed to model the scheduling and usage of the FCRs for reconfiguring the control center for flight operations.

The problem arises when several flights, in different stages of preparation, require the same FCR and these demands temporarily exceed the capacity of the system. To some extent the resource contention can be dealt with by management, but eventually, with increasing frequency of flights, this becomes increasingly complex, if not impossible, and an additional FCR will be required. Because of the tremendous cost and long lead time required to build a new control room, a model of the scheduling of the existing control rooms was necessary. This model is used to find out how to get by with the two FCRs as long as possible and to predict when a third room will be needed.

APPROACH

The approach chosen to solve the scheduling and resource problem was to employ the "expert system" or "production system" concept [Buchanan82], [Stefik82]. The production system used is called the Rule Based Modeling System (RBMS) developed by Ford Aerospace & Communications Corporation. The RBMS consists of

the following:

- An Inference Engine
- A Report Generator
- A Rule file
- A LISP function file

For an illustration of the functional flow of RBMS, see Figure 1.

The overall strategy of the RBMS is to develop a set of rules and LISP functions that schedule the FCR the same way that it is scheduled by the human "experts". The sequence of events that occurs during an analysis is as follows:

- A rule-set for application is loaded into working memory.
- Associated supporting LISP functions for the rules are loaded into memory.
- The flight manifest is loaded into working memory.
- Control is handed over to the Inference Engine.

At this time the inference engine assumes full control over RBMS and performs the inference task as dictated by the loaded rule-set. The particular inference engine used by RBMS is OPS-5, developed by C. L. Forgy at Carnegie-Mellon University [Forgy81], [McDermott82]. When no further inferences may be achieved by the inference engine, the schedule can be printed and post processing programs can be run to generate reports for the user.

SCHEDULING RULES

The process for determining how the FCRs will be scheduled is contained in the rule-set that is resident in the rule file. This file consist of rules of the "If...then..." form, sometimes called "situation-action". The situation ("if" portion) of the rule is a condition or state of the problem. The inference system determines if the condition or state of the "situation" is satisfied. If the "situation" is satisfied, then the rule "fires" and the specified "action" ("then" portion) takes place. If the situation of a rule is only partially satisfied, then no action takes place. The rule in Figure 2 is an example of a situation-action rule.

The approach that was used to build the rule file was to question the "experts" on how the FCRs were being scheduled and then represent each of these scheduling guidelines with an OPS-5 rule and special purpose LISP functions. These scheduling guidelines may have unique constraints which might otherwise prove difficult to represent using conventional software techniques.

"""

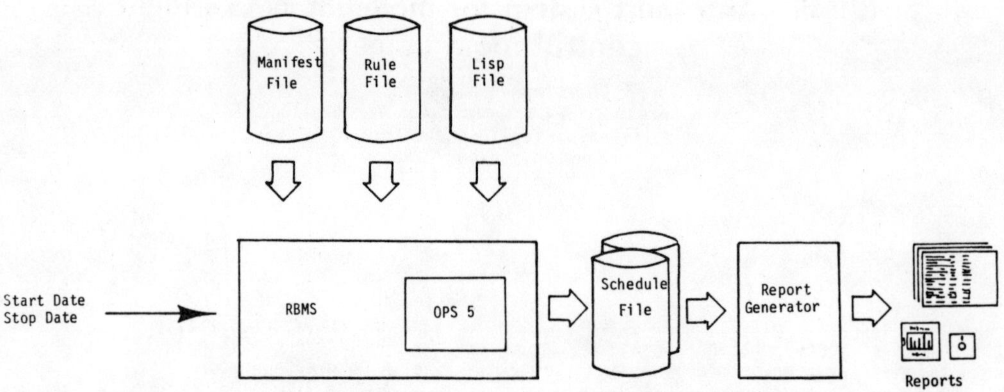

Figure 1. RBMS System Flow

```
(p schedule-pocc
     (schedule-rest)
     (flight  name  n   start-date  sd   s-level
        =2  room  r  )

     (call passit6  sd  84 1 7 4  r  )
     (bind  date  (schedule-no-sunday))
     (write  schedule  date  (tabto 8)  r  (tabto
        10)  n  (tabto 20)  pocc-test  (tabto 31)
           4  (crlf))
     (modify 2  s-level 3))
```

English equivalent:

If it is time to schedule the Payload Operations Control Center (POCC) and there is a flight that needs to
have the POCC scheduled, then schedule it 84 days
before launch within a 7 day window for 4 hours but
not on Sunday. Write that information to the schedule
file and mark the flight as being scheduled for the
POCC.

Figure 2. Rule Schedule-POCC

The following assumptions were used in modeling
the FCR usage:

- FCR usage would take place on three shifts
  per day, seven days a week.

- All DOD flights will be assigned to FCR #2.

- All Spacelab flights will be assigned to
  FCR #1.

- Other flights will be assigned to give a
  balanced FCR loading.

- Some activities may not take place on Sundays.

The following scheduling guidelines were established for modeling FCR usage:

- STS flights take one FCR from two days prior
  to launch to 12 hours after launch and are
  fixed.

- Two 48 hours periods of power outage are
  scheduled each year on weekends and take
  down all FCRs.

- One day per quarter is used for equipment
  installation on each FCR.

- Maintenance is scheduled in six four hour
  blocks a week on any FCR that has time
  available.

- Systems integration testing is scheduled
  for one five hour block per week for six
  weeks twice a year.

- Based on the complexity of the flight,
  simulation will be scheduled for high
  complexity or for low complexity.

- The high complexity simulations will be
  scheduled for 10 hours a day twice a week
  for eight weeks.

- The low complexity simulations will be
  scheduled for 10 hours a day twice a
  week for six weeks.

- All simulations must be scheduled within
  a seven day window.

- Three hours of time are required for preparation of each simulation.

- Pad test will take place for two eight hour periods two weeks prior to launch within a three day window.

- Network simulations take place two weeks after the first simulation for eight hours within a 14 day window and three weeks prior to the flight for eight hours within a seven day window.

- Interface testing between the MCC and the simulator take place prior to the first week of sims within a 14 day window, and three weeks prior to a flight within a seven day window for five hours each time.

- Payload Control Center testing takes place during the first week of simulations in three four hour blocks within a 14 day window.

The activities are scheduled in the following order:

- Real flights are scheduled first.

- Power outage.

- Equipment installation.

- Simulations and all other pre-flight specific activities.

- Maintenance and systems integration is scheduled last.

After running the model on a given flight manifest, the results can then be analyzed. Changes to the way that the FCRs are scheduled can be incorporated into the model by changing the rules and running the program. Once a given rule set is established it is easy to see the effect of changing the flight manifest on FCR usage.

RESULTS

The RBMS has proved to be extremely useful for modeling FCR usage. Given the typical flight manifest shown in Figure 3, a schedule like the one in Figure 4 will be generated using this system. Further post-processing can then take this output and put it into a standard spread sheet on a PC where further time, changes to the manifest and rules can be made, the program run, and results generated for comparison with the original run. (See Figure 5) This program was so useful in modeling FCR usage that it is now also being used to look at other resources in the Mission Control Center.

SUMMARY

The RBMS has been used to evaluate the flight load rate of the Shuttle Transportation System and identify potential problems in the scheduling of the Flight Control Rooms. It is particularly useful in evaluating the cost effectiveness of new procedures and for planning future missions.

An advantage that the production system concept has over more conventional approaches is that the algorithm one chooses can be more easily expressed by the human expert. The user may more readily experiment with heuristic concepts in the interactive environment of the production system.

FLIGHT NAME	LAUNCH DATE	SIMULATION COMPLEXITY	NASA/DOD	DURATION	SPACELAB	FCR #
STS-51A	84307	l	NASA	6	n	1
STS-51C	84345	m	DOD	7	n	2
STS-51B	85017	m	NASA	7	y	1
STS-51E	85043	l	NASA	5	n	2
STS-51D	85077	l	NASA	6	n	1
STS-51F	85107	m	NASA	7	y	2
STS-51G	85150	l	NASA	7	n	1
STS-51L	85183	l	NASA	7	n	2
STS-51I	85220	l	NASA	7	n	1
STS-51J	85263	m	DOD	7	n	2
STS-61A	85287	m	NASA	7	y	1
STS-62A	85288	m	DOD	7	n	2
STS-61B	85310	l	NASA	7	n	1
STS-51H	85331	l	NASA	7	n	1
STS-61C	85358	l	NASA	7	n	2
STS-61D	86022	m	NASA	7	y	1
STS-61E	86065	l	NASA	7	n	2
STS-61F	86135	m	NASA	2	n	1
STS-61G	86141	m	NASA	2	n	2
STS-61H	86178	l	NASA	7	n	1
STS-61I	86197	l	NASA	7	n	2
STS-61J	86225	m	NASA	3	n	1
STS-61K	86233	m	DOD	7	n	2
STS-61L	86255	l	NASA	7	n	1
STS-62B	86272	m	DOD	7	n	2
STS-71A	86274	l	NASA	7	n	1
STS-71B	86296	l	NASA	7	n	1
STS-71C	86310	m	DOD	7	n	2
STS-71D	86330	l	NASA	7	n	1
STS-71E	86352	l	NASA	7	n	2

Figure 3. Manifest File

DATE	FCR #	FLIGHT	ACTIVITY	DURATION (HOURS)
86001	1	STS-61D	mcc-sms	5
86001	2	none	IV-maint	4
86002	1	STS-61D	SIMs	13
86002	2	STS-61E	re-test	1
86003	1	none	IV-maint	4
86003	2	none	IV-maint	4
86004	1	STS-61D	SIMs	13
86004	2	none	IV-maint	4
86006	1	none	IV-maint	4
86006	2	none	IV-maint	4
86007	1	none	IV-maint	4
86007	2	none	IV-maint	4
86008	1	STS-61D	N-vals	16
86008	2	none	IV-maint	4
86009	1	none	IV-maint	4
86009	2	STS-61E	SIMs	13
86013	1	none	IV-maint	4
86013	2	STS-61E	SIMs	13
86014	1	none	IV-maint	4
86014	2	none	IV-maint	4
86015	1	none	IV-maint	4
86015	2	none	IV-maint	4
86016	2	STS-61E	SIMs	13
86017	2	none	IV-maint	4
86020	1	STS-61D	FCR-loc-pr	24
86020	1	none	IV-maint	4
86020	2	STS-61E	SIMs	13
86021	1	STS-61D	FCR-loc-pr	24
86021	1	none	IV-maint	4
86021	2	none	IV-maint	4
86022	1	STS-61D	Flight	24

Figure 4. Schedule File

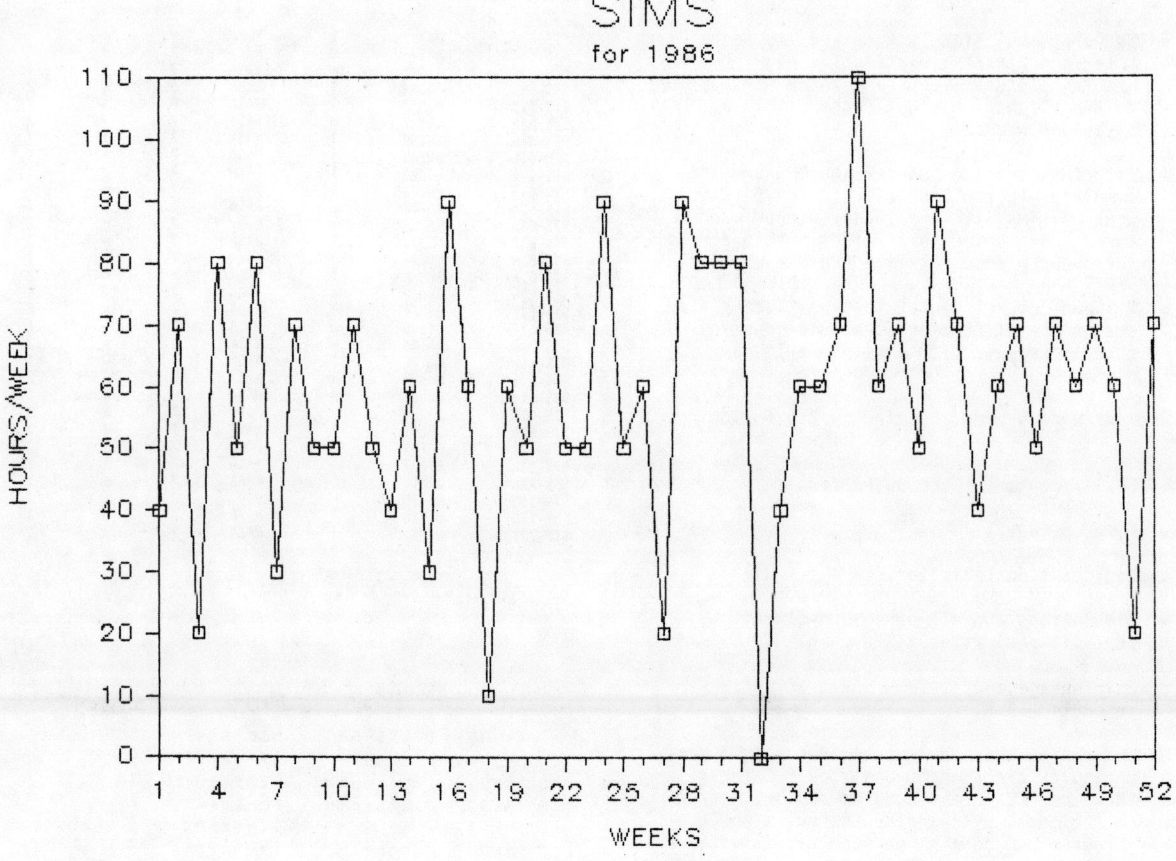

Figure 5.

REFERENCES

[Buchanan82] Buchanan, B. G. and R. Duda, "Principles
    of Rule-Based Expert Systems", HPP-82-14,
    Stanford University, August 1982.

[Forgy81] Forgy, C. L. "OPS5 Users Manual", Carnegie-
    Mellon University, Pittsburg, Penn., 1981.

[McDermott82] McDermott, J., "R1: A Rule-based Con-
    figurer of Computer Systems", Artificial
    Intelligence, 19, (1982).

[Stefik82] Stefik, M., J. Aikins, R. Blazer,
    J. Benoit, L. Birndaum, F. Hayes-Roth, and
    E. Sacerdoti, "The Organization of Expert
    Systems: A Tutorial", Artificial Intelligence,
    18 (1982).

Simulating aphasia

William E. Leigh
University of Southern Mississippi
Hattiesburg, Mississippi

Noemi M. Paz
University of Southern Mississippi
Hattiesburg, Mississippi

H. Ray Souder
Northern Kentucky University
Highland Heights, Ky.

ABSTRACT

Aphasia refers to the pathology of language and speech resulting from head injury. The selective disruption of the language system in the abilities of lexical-semantic understanding or syntactic discernment is established. We are experimenting with an optimization model of the language understanding process which has submodels for syntax and semantics. This model when perturbed exhibits symptoms which may be interpreted as aphasic.

WEPMAN'S THEORY OF APHASIA

Joseph Wepman [1] has designed a model of language disturbance which provides for three levels of function (conceptual, perceptual, and reflexive) and three processes (input transmission, integration, and output transmission). We confine our work to the perceptual function and the process of integration. Wepman proposes a five-category hierarchy of severity of aphasia which he believes represents the stages of recovery as well as impairment. These categories are (1) syntactic aphasia, (2) semantic aphasia, (3) pragmatic aphasia, (4) jargon aphasia, and (5) global aphasia. We are interested in syntactic aphasia, in which the patient is unable to make use of his previous knowledge of grammatical structure, and semantic aphasia, in which the patient has difficulty in using names and labels correctly.

AN OPTIMIZATION MODEL OF LANGUAGE UNDERSTANDING

In previous work [2] we have developed an optimization model of natural language database query interpretation. This model involves developing a 0-1 integer programming problem for each query. The possible values that the set of variables in the optimization may take on is an enumeration of possible individual associations between the database schema and the query. The solution to the integer programming problem corresponds to a single assignment of database data items and relationships to the words in the query. Constraints are derived from syntactic and semantic knowledge stored as libraries of templates. An objective function is used to rank the possible associations as to their likelihood of agreement with the intent of the questioner.

Syntax and semantics are used as sources for constraints. Interpretation is the process of solving the model. Ambiguity is a quality of the query if alternate optimal solutions are possible. The necessary objective function is a result of experimentation. Coupling constraints are used to tie together the submodels, which are (1) the syntactically derived constraints and (2) the semantically derived constraints.

PERTURBING THE MODEL

This model was tested on 400 database queries. Variations of the model were also tested. The model in its entirety performed in what may be termed a "normal" manner. The model, when varied by selectively removing some of its function, behaved in a degraded manner. This degradation was of a nature which may be considered to be "aphasic."

When the syntax derived constraint submodel was removed, the system processed twice as many queries incorrectly. However, due to the nature of the solution process, the processing of the queries with the reduced model took much less time. The reduction of the semantic knowledge in the system also reduced the processing time. It was found that this knowledge could be removed in such a way as not to increase incorrect interpretations. If this knowledge were removed randomly, it would increase incorrect interpretations.

INTERPRETING THE MODEL'S BEHAVIOR

The experience with this model suggests that optimization may be a useful modelling framework for normal and pathological language understanding. The impact on performance caused by removing the syntax submodel is what would be expected from applying Wepman's hierarchy of impairment. The impact of larger amounts of knowledge on processing time suggests that aphasia might also result from brain processing impairment as well as knowledge removal.

Experience with this model leads us to believe that database querying is a domain of language use which operates with a restricted pragmatics which makes syntactic knowledge more redundant and less crucial than it may be in other domains. These results may be corroborated in experimental work with aphasics. We are searching for this type of data to guide us in further experiments.

REFERENCES

[1] Wepman, Joseph, RECOVERY FROM APHASIA, Ronald, New York, 1951.

[2] Leigh, William, and Souder, Ray, "The Interpretation of Natural Language Database Queries with Optimization Methods," PROCEEDINGS OF CONFERENCE ON ARTIFICIAL INTELLIGENCE, April 26-27, 1983, Oakland University, Rochester, Michigan.

Assessment of proportions: An efficient method for single neuron recognition of letters, numbers, faces and certain types of concepts

Rolf Martin
Chemistry Department
Brooklyn College
Brooklyn, NY 11210

ABSTRACT

Many concepts and concrete entities may be distinguished from similar ideas and objects by examination of key-attribute proportions. For example: giraffes and antelopes can be differentiated by the ratio or proportion of neck and body length. (Neck length alone is unimportant since giraffes and antelopes are not confused when the neck of a distant giraffe appears to be shorter than that of a much closer antelope.) Similarly, human faces, written characters and many other patterns may be distinguished from one another, regardless of their absolute size, by the relative proportions of horizontal, vertical and diagonal lines, angles, and line endings. In this paper, two learning rules are shown to enable individual network elements to recognize attribute proportions: 1) connections between network elements are strengthened when two or more discharge concurrently (Hebb's rule), and 2) the combined strength of the connections leading to each network element tends to reach a similar total value. The ability to recognize and respond to proportions results from the interaction of these two learning rules with an algorithm for summation of input to each network element that corresponds to the formation of neurotransmitter-receptor complexes such as those found in typical nerve networks. This explanation for proportion assessment is of interest because it enables single elements or cells to recognize typical or deformed, upright or rotated, shifted or centered, reduced in size or expanded, script or printed, uppercase or lowercase letters, numeric characters, faces and other patterns within a network that is only 1.5% to 5% the size of comparable nets described elsewhere. The two learning rules are of interest because they are a subset of six that have previously been shown to provide a basis for concept representation, learning by simile, risk benefit assessment, game playing and information matching and sorting.

MAJOR SECTION HEADINGS

Introduction
Network Structure
Recognition of Proportions
Mechanisms of Learning
Recognition of Letters, Numbers and Faces
  Training the network
  Testing the network
  Comparison with Fukushima's Neocognitron
  Decreased response to deformed faces
Discussion
  Evaluation of efficiency, accuracy and errors
  Key features and concepts

INTRODUCTION

Recognition of letters and numbers apparently depends on recognition of key features (1), whether they be geometric features, pattern frequencies, or both, and may also depend on accurate assessment of the relative proportions of key features. For example, a script "e" and "l" may be distinguished by assessment of the ratio of height to width; by the ratio or relative proportions of vertical, near vertical, diagonal, horizontal and near-horizontal lines; or by pattern frequency analysis that includes evaluation of relative peak heights.

Certain concepts are also identified or defined by relative proportions. Giraffes and antelopes, rabbits and guinea pigs, geese and ducks, squares and rectangles, circles and ovals, rods and cylinders, spots and blotches, stripes and bands, cars and limousines, and many others are differentiated by key attribute proportions that are evaluated by the visual system. Foods are identified in major part by the relative amounts of sweet, bitter and salty ingredients, as well as other scents and flavors. And sounds are very often categorized or identified by the relative volumes or proportions of different vibrational frequencies. So it appears that the ability to assess relative proportions of different sensory inputs may be a key component of neuronal information processing, and may also provide a useful basis for artificial pattern recognition.

In this paper a simple mechanism for recognition of proportions is described and shown to enable neuron-like networks to recognize a variety of input patterns, including letters, numbers and faces of the types shown in Figure 1. This model of pattern recognition is characterized by: 1) only one layer of variable-strength connections - from network elements that function as feature detectors to those that recognize and respond to particular letters or patterns; 2) rapid adjustment of these connections to the required values, so that relatively accurate recognition of faces, numbers and letters can occur after the first exposure to one example of each character or pattern; and 3) rapid information processing that includes comparison during each loop of computation of each feature of the current pattern with the corresponding features of all previously learned patterns, and calculation of similarity ratings that enable an immediate decision concerning the identity of the current pattern. Use of a single layer of variable-strength connections is critical since it removes from consideration questions of "credit assignment" (2) that have plagued multi-step or hierarchical network models for decades.

Despite some initial successes, however, the approach taken appears to be some distance from a useful (in a practical or commercial sense) system for character recognition. The aim of this paper is primarily to explore the properties of network elements or neurons that function as proportion detectors.

Recognition of proportions was simulated in a three layer network of 400 retinal cells, 54 primitive feature detectors and 120 proportion detectors that recognize and signal the identity of each input pattern. This structure, which is shown in Figures 2 and 3, has fewer layers of network elements than related models, and fewer elements in each layer. The "Neocognitron" of Fukushima et al. (3,4), for example, contains nine layers and approximately 15,000 elements, while the net of Marko and Giebel (5,6) is (apparently) composed of six layers and more than 8,000 cells.

The network elements used here are explicit models of neurons: during simulations they release and bind neurotransmitters, experience fatigue and lateral inhibition, are connected by asymmetric links, and fire at variable frequencies (after a threshold has been reached) according to the total input each element receives from others. The constraint on the total strength of connections (synaptic weights) to each element (learning rule 2) is intended to reflect a physiological limit on the total conductivity of a neuron's dendritic tree.

RECOGNITION OF PROPORTIONS

As a consequence of the learning rules noted above, individual network elements (henceforth "neurons") are able to measure the extent to which the proportion of inputs received from other neurons resembles previous input proportions. For example: if neuron A has received stimulation from neurons (feature detectors) FD1, FD2 and FD3, and the amounts of stimulation from each were proportionately 1 to 2 to 7 (henceforth "1:2:7"), then neuron A will become maximally sensitive to future inputs from FD1, FD2 and FD3 that are proportionately 1:2:7. Thus if these three feature detectors are (directly or indirectly) triggered by horizontal, vertical and diagonal lines, respectively, then neuron A will respond maximally to (and hence will recognize) patterns that contain the particular combination of horizontal, vertical and diagonal lines that triggers a 1:2:7 discharge of FD1, FD2 and FD3. Input of patterns that result in 1:3:6 and 2:2:6 firing of the three feature detectors will lead to a less vigorous response by neuron A.

Let us postpone for a moment consideration of how neurons can learn to respond to particular input proportions and examine instead the mechanisms by which they may respond selectively once learning has been completed. Consider the network in Figure 2, which contains the neurons of the previous paragraph and two additional neurons, B and C, that are most effectively triggered by 2:2:6 and 1:3:6 inputs from FD1, FD2 and FD3.

The input pathways to (henceforth "dendrites of") neurons A, B and C each contain receptors which bind neurotransmitter molecules released from the output pathways ("axonal branches") of the feature detectors. If the degree of stimulation to each neuron is a function of the number of neurotransmitter-receptor complexes formed in all of the synapses leading to the neuron, if all neurons produce and maintain a similar number of receptors, and if the number of neurotransmitters released in response to input patterns remains relatively constant, then each input pattern will stimulate most effectively (and thus be recognized by) neurons that have in effect "placed" their receptors to <u>match</u> the relative number of neurotransmitters discharged into each synapse.

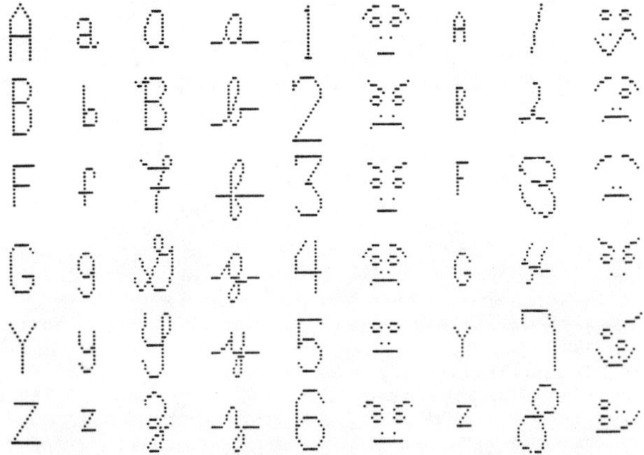

<u>FIGURE 1</u>: Letters, numbers and faces recognized by the network. Each of these patterns, and others, was imposed on a 20 x 20 grid of retinal cells and then transmitted to other neurons that detect features and feature proportions. One hundred and twenty patterns of the types shown in columns 1 to 6 above were used as typical examples or "template" patterns to train neurons that detect feature proportions to recognize letters, numbers and faces. After this training, test patterns were input to determine whether these proportion detectors could serve as accurate pattern identification cells. Test patterns included those in columns 1 to 6 that were shifted to the right and left, letters of reduced size (column 7), rotated or deformed numbers (column 8) designed and originally used by Fukushima, Miyake and Ito (3), and deformed faces.

Such recognition is illustrated in Figure 2. Note that the relationship discussed in Figure 2 and Table 1 between the number of transmitter molecules released and the number of neurotransmitter-receptor complexes formed is a key aspect of this model. If transmitter and receptor concentrations are simply multiplied when the degree of stimulation to each neuron is calculated, then neurons will not accurately recognize input proportions.

MECHANISMS OF LEARNING

The conductivity of each synapse can be adjusted to match input proportions 1) if the capacity to release neurotransmitters from each presynaptic axon terminal increases in relation to the discharge frequency of the presynaptic (feature-detecting) neuron whenever both pre- and postsynaptic neurons discharge concurrently, or 2) if the amount of input received at each synapse determines, or is correlated with mechanisms that determine, the number of receptors at each synapse. Many molecular mechanisms are possible: all that is required is that the strength of each connection be regulated according to firing activity, much as sunlight determines the proportionate size of each branch on a tree. One possibility is that transmitter-receptor complexes are more stable (less easily degraded) during periods of neuronal discharge than are empty receptors, as if transmitters bind to and protect their receptors (8). As a result, "excess" receptors (those <u>not</u> occupied by transmitters when firing occurs) will be removed when each neuron is triggered.

Pattern recognition was carried out in three steps. The image of a letter or face to be learned was first entered onto a retina by mechanisms that are not part of the network simulation. The retinal image was then transformed, by lateral inhibition involving neighboring cells only, into a 'first derivative' image of the kind shown in Figure 3. This 'first derivative' corresponds to the output of the retinal layer to a group of feature detectors that are triggered by (and hence detect) horizontal, vertical and diagonal lines, areas of excitation subject to low lateral inhibition (line endings) and "silent" areas subject to high lateral inhibition (sharp bends or line intersections). Separate sets of these feature detectors receive input from the upper, lower, right and left halves of the image, from each of four quadrants, and from the entire image, so that the occurrence of each feature in different areas of the image can be recognized. Pattern-identification neurons (identification or recognition cells corresponding to neurons A, B and C of Figure 2) tally the input they receive from the feature detectors and then discharge if input proportions match those of the character or pattern each represents.

Training the network

To give the character-identification neurons (proportion detectors) an "idea" of the features each must recognize (i.e. to adjust the receptor levels of each identification cell to enable recognition of the correct feature proportions), a single, typical example of each letter, number or face was placed on the

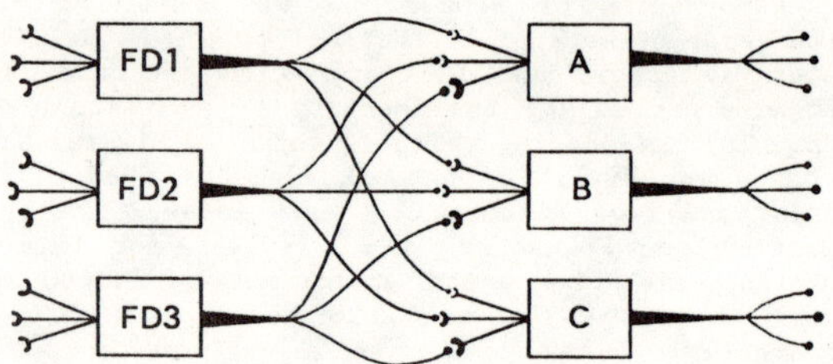

FIGURE 2: Recognition of proportions. In this illustration, neurons A, B and C each have a total of 5,000 receptors distributed among synapses in the same proportion as the optimal stimulus pattern for each neuron. If neurons FD1, FD2 and FD3 release 1,000, 2,000 and 7,000 neurotransmitter molecules, respectively, into the synapses that each forms with A, B and C, then neuron A will respond most vigorously since this input matches the 1:2:7 apportionment of receptors within synapses leading to (stimulating) neuron A. Table 1 provides a breakdown of transmitter-receptor binding within each synapse and the resulting stimulation received by each neuron. If on the other hand FD1, FD2 and FD3 release 2,000, 2,000 and 6,000 transmitters, or 1,000, 3,000 and 6,000 transmitters, then by the mechanisms outlined in Table 1 either neuron B or neuron C will fire most vigorously.

TABLE 1: Neurotransmitter-receptor interactions that enable recognition of proportions. In this table, FD1, FD2, FD3 and the letters A through C refer to neurons in Figure 2. In this example the number of transmitter-receptor complexes that are formed never exceeds the number of receptors in any synapse, and half of the transmitters released into each synapse are degraded or removed from the synapse by reuptake mechanisms and thus never bind to receptors. Consequently the number of transmitters released into the synapse from FD2 to C is insufficient to fill each available receptor and the number of transmitter-receptor complexes formed in synapses from FD3 to B and FD3 to C is less than half the number of transmitters released into each of these synapses. As a result, synapses leading to neuron A contain the greatest number of transmitter-receptor complexes and neuron A fires most vigorously, as if to identify the 1:2:7 input ratio.

Synapse	Number of neurotransmitters released into each synapse	Number of receptors in the synapse	Number of transmitter-receptor complexes formed	Total number of complexes stimulating each neuron
from FD1 to A	1,000	500	500	
from FD2 to A	2,000	1,000	1,000	
from FD3 to A	7,000	3,500	3,500	5,000 (to neuron A)
from FD1 to B	1,000	1,000	500	
from FD2 to B	2,000	1,000	1,000	
from FD3 to B	7,000	3,000	3,000	4,500 (to neuron B)
from FD1 to C	1,000	500	500	
from FD2 to C	2,000	1,500	1,000	
from FD3 to C	7,000	3,000	3,000	4,500 (to neuron C)

```
| 0004444000 |             | 0004444000 |             | 5422305522 |             | 321 316 340 |
| 0040000400 |  Step 1     | 0040000500 |  Step 2     | 2303420006 |  Step 3     | 306 481 364 |
| 0040000000 | ----------> | 0040000000 | ----------> | 2002303400 | ----------> | 341 287 365 |
| 0040000400 |  lateral    | 0040000500 |  feature    | 2002120309 |  proportion | 324 377 332 |
| 0004444000 |  inhibition | 0004444000 |  analysis   | 6822004124 |  analysis   | 423 304 298 |

    Retinal                   First                    Second              Output firing rates of
    image                     derivative               derivative            identification cells
                                                                            (only 15 of 120 are shown)
```

FIGURE 3: Overview of the steps employed during pattern recognition. A retinal image, similar to that shown above, is first imposed on the firing frequencies of 400 retinal neurons. Those that are excited by the input pattern discharge at an arbitrary frequency of 4; those elsewhere are assigned firing rates of zero. A first derivative image is then created as a result of lateral inhibition involving neighboring cells. Areas of excitation and low lateral inhibition (such as line endings) are assigned a firing frequency of 5 units. A second derivative is composed of the firing rates of 54 feature-detecting cells that respond to 6 features in each of nine regions of the retina. Output from these feature detectors stimulates 120 proportion-detecting cells that respond most vigorously to, and hence identify, previously encountered feature ratios. Each proportion detector or "identification" cell receives input from every feature detector in the previous layer and recognizes relatively specific input proportions as described in the text. Typical output firing rates of identification cells are shown in Figure 4.

retina. At the same time, one and only one of the proportion detectors was triggered -- as if in response to separate auditory input of the name of the letter appearing on the retina. Concurrent discharge of this proportion detector and the feature detectors triggered by retinal input strengthens the connections between them, as first postulated by Hebb (9). Proportions are represented by the relative strength of connections leading to the proportion detector because each feature detector discharges at a frequency that reflects the number of occurrences of the feature it recognizes. Such discharge then results in adjustment of synaptic conductivities to levels that reflect the number of occurrences of each feature. Further details of the simulation procedure, including program code with comments and a discussion of the related capabilities of this type of network, have been published recently (7,10).

Testing the network

After examples of each of four sets of letters (upper and lowercase, printed and script), ten numbers (0 through 9) and six faces were input into the network during training, each character or face was input once again to test whether identification cells would respond maximally to the character each was "trained" to recognize. No recognition errors were made in response to the 120 original (henceforth "template") patterns.

Additional "test" patterns, not previously seen by the network, were then input to determine the extent to which this net could "apply" its "knowledge" to novel patterns. Samples of the test patterns are shown in Figure 1. No recognition errors were made when patterns were simply shifted one column to the right or left, or when slightly distorted and smaller characters and faces were used.

More severe distortions of course resulted in recognition errors, more or less according to the severity of the deformations. When only a single

template of each character had been learned, and the network had no opportunity to learn which features of each character were defining or essential, character recognition appeared to be comparable to, but somewhat less accurate than that achieved by Fukushima et al. (3), who designed a nine-layer network of approximately 15,000 neural elements and trained it with an average of four templates for each of the numbers 0 through 9.

Comparison with Fukushima's Neocognitron

To compare the two approaches more carefully, the deformed test numbers designed by Fukushima et al. (3) were copied and presented to the net of 54 feature detectors and 120 identification cells described in Figures 2, 3 and 4. This net was able to recognize 40% of the deformed numbers, noticeably less than the 60% success rate achieved by Fukushima's Neocognitron.

To increase the accuracy of the identification cells, several approaches were considered: 1) train each identification cell with more than one template so that each learns to respond selectively to the defining features of a particular character; 2) train each identification cell with one template only, but add to the network large numbers of additional identification cells that each recognize a single, typical deformation of a particular character; 3) add additional learning rules that will result in formation of inhibitory connections that block inappropriate responses and/or weaken excitatory connections that cause errors; and 4) add key-feature detectors governed by additional learning rules to a net of proportion detectors that are unaffected by the additional rules, and work with a heterogeneous net composed of both types of cells.

Option (1) was rejected because the author did not wish to become entangled in endless fiddling with the precise shapes of hundreds of templates (far better to enable the net to select, on its own, fundamental templates from among those it has encountered), and because alteration of synaptic weights to enable identification cells to respond selectively to defining or key features appears inevitably to destroy their ability to recognize feature proportions, an unacceptable sacrifice. Option (2) was pursued since addition of large numbers of additional identification cells to such a small net could be accomplished quite easily and because the formation of potentially useful "committees" of these identification cells was anticipated. And options 3 and 4 are being held for the future, to be pursued when exploration of the simplest, homogeneous nets of proportion detectors has progressed further.

Training greater numbers of identification cells with additional templates, including examples of numbers that were shifted, rotated and of similar general style as those of Fukushima et al., greatly improved the accuracy of the proportion detectors. Correct responses were obtained for 77% of the deformed numbers. Examples of those that were properly identified are shown in Figure 5; those not recognized are shown in Figure 6. Additional training, with templates resembling characters that were not recognized, could in principle increase the accuracy of proportion detectors further.

Decreased Response to Deformed Faces

The relationship between degree of pattern distortion and frequency of identification cell discharge was most clearly evident when test patterns were deformed faces. Although such patterns were invariably recognized as faces, the response of proportion detectors trained to identify them decreased almost linearly with the amount of deformation. When one eye was omitted from the test pattern, for example, the response was generally about 10% less than the response when both eyes were present, and 10% greater than when both eyes were missing. Appropriately, the response to the most severely distorted face (shown in Figure 1 at the bottom of column 9) was even less vigorous. This type of response is similar to that of neurons which are selectively triggered by patterns resembling a monkey's hand: other patterns also trigger these "hand recognition" cells, however the strongest responses are obtained when patterns are very similar to a monkey's hand (11,12).

Thus, despite some assertions to the contrary (13,14), there seems to be no reason to rule out the existence, in either animals or humans, of millions of partially-specific, pattern-detecting neurons that each respond most strongly to a single pattern, concept or subset of attributes, with a specificity determined as much by inhibition among neighboring cells as by the pattern of strengthened connections to each. Interested readers are referred to pp. 42-44 of reference 2 for a more detailed consideration of neuron specificity.

DISCUSSION

Evaluation of efficiency, accuracy and errors

Quantitative evaluation of the efficiency of a character recognition system is an almost impossible task. How does one weight the relative importance of network size, number of template patterns required, and time and space requirements for each character, 100 characters, and 1,000 characters? Evaluation of the accuracy of a recognition mechanism is just as difficult. Statements concerning the percentage of correct responses have relatively little meaning

ID cell no.:	1	2	3	4	5	6	7	8
template:	A	a	B	b	C	c	D	d
firing rate:	321	316	340	306	481	364	341	287

ID cell no.:	9	10	11	12	13	14	15	16
template:	E	e	F	f	G	g	H	h
firing rate:	365	324	377	332	423	304	298	281

FIGURE 4: Firing frequencies of the first 16 identification cells after input of a printed, uppercase "C". The identification neuron with the maximum firing rate (481) is that which had previously been trained to respond to a sample "C". Each of the other identification cells responds with an intensity that reflects the similarity of the present pattern to the template pattern previously used to train the neuron. Note that the cell previously trained to recognize "G" has the second most vigorous response. This is appropriate since an uppercase "G" closely resembles the current input pattern "C". As described in (7), firing rates such as these may correspond to four types of similarity ratings, including two that bear some resemblance to "selfish" and "selfless" similarity assessments.

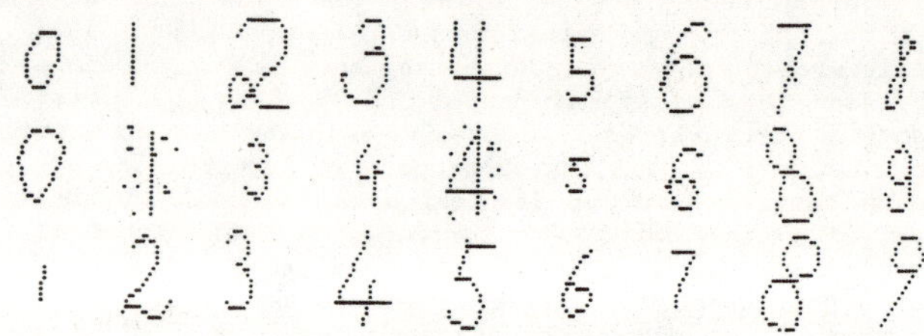

FIGURE 5: Deformed test patterns that were correctly identified by the net of feature and proportion detectors. Each of these test patterns, and those of Figure 6, was designed by Fukushima et al. (3). Six additional patterns that were correctly recognized are shown in column 8 of Figure 1.

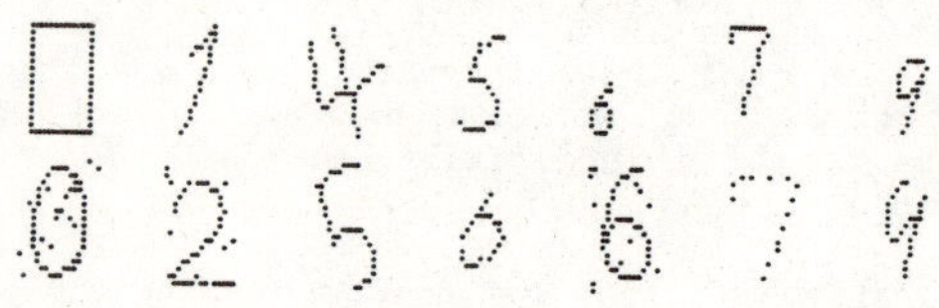

FIGURE 6: Examples of test patterns that were incorrectly identified. Several of the errors involve numbers that are similar to letters (e.g. "2" and "Z", zero and "o", "5" and "S"). Other errors result from the effects of noise (random gaps or dots that are mistaken for line endings). These errors suggest that the network of feature and proportion detectors requires significant modification, perhaps to enable evaluation of additional or defining features.

unless the degree of test character distortion is specified. Unfortunately mathematical measures of the "distance" between template and test characters do not convey critical information about key features that are the most important for accurate identification.

For measurement of progress and comparison of alternative approaches, a standard set of test characters appears to be required. This paper has followed the lead of Fukushima, Miyake and Ito (3), who published examples of test patterns that could and could not be recognized by their net. The deformed characters they designed were again used as test patterns, and will continue to serve in that role in future simulations at Brooklyn College. It would be interesting to see other character recognition systems tested with these patterns as well.

Key features and concepts

Proportion detectors of the type used in the simulations reported here, which receive and recognize information about all detectable features from each region of the input retina, cannot represent or evaluate defining or key features because adjustment of synaptic conductivities to reflect the importance of such features impairs the ability of these cells to recognize input proportions. Key features can, however, be represented and evaluated by nets containing less-complex proportion detectors that receive information about a limited number of features from one or two areas of the retina and then signal the presence of particular ratios of these features to neurons that function as concept identifiers. The importance of each key-feature ratio can then be represented as an increase in the conductivity of the synapse linking the simple proportion detector to the corresponding concept recognition cell, without impairing the ability of either cell to recognize proportions or key features. Preliminary simulations suggest that feature ratios or proportions can then be treated as more-or-less ordinary attributes of concepts represented in the net.

ACKNOWLEDGEMENT

The author gratefully acknowledges the assistance of Aaron Lukton, Roy D. Martin, John A. Evans, George Emmanouilidis, Anthony Stergianopoulos and Anacletos T. Padua during preparation of this paper, and the computer facilities and services provided by the Chemistry Department and the Computer Center at Brooklyn College.

LITERATURE CITED

1) M.J. Gervais, L.O. Harvey, Jr. and J.O. Roberts, "Identification Confusions Among Letters of the Alphabet," J. Exptl. Psychol.: Human Perception and Performance 10(5), 1984, 655-666.

2) J.A. Anderson and G.E. Hinton, "Models of Information Processing in the Brain," in "Parallel Models of Associative Memory," J.A. Anderson and G.E. Hinton, Eds., Erlbaum, Hillsdale, New Jersey, 1981, 9-48 (p. 15).

3) K. Fukushima, S. Miyake and T. Ito, "Neocognitron: A Neural Network Model for a Mechanism of Visual Pattern Recognition," IEEE Trans. on Systems, Man and Cybernetics, SMC-13(5), 1983, 826-834.

4) K. Fukushima, "Neocognitron: a Self-Organizing Neural Network Model for a Mechanism of Pattern Recognition Unaffected by Shift in Position," Biological Cybernetics 36, 1980, 193-202.

5) H. Giebel, "Feature Extraction and Recognition of Handwritten Characters by Homogeneous Layers," in Pattern Recognition in Biological and Technical Systems, O.J. Trusser and R. Klinke, Eds., Springer, New York, 1971, 162-169.

6) H. Marko, "A Biological Approach to Pattern Recognition," IEEE Trans. on Systems, Man and Cybernetics, SMC-4(1), 1974, 34-39.

7) R. Martin, A. Lukton and S.N. Salthe, "Simulation of Simple Cognitive Maps, Concept Hierarchies, Learning by Simile, and Similarity Assessment in Homogeneous Neural Nets," Proceedings of the 1984 Summer Computer Simulation Conference, Society for Computer Simulation, vol. 2, 1984, 808-821.

8) G.S. Stent, "A Physiological Mechanism for Hebb's Postulate of Learning," Proc. Natl. Acad. Sci. (US) 70(4), 1973, 997-1001.

9) D.O. Hebb, "The Organization of Behavior," Wiley, New York, 1949.

10) R. Martin, "A Possible Mechanism for Neuronal Representation of Concept Hierarchies," Proceedings of the 28th Annual Meeting of the Society for General Systems Research, vol. II, 1984, 507-510.

11) C.G. Gross, C.E. Rocha-Miranda and D.B. Bender, "Visual Properties of Neurons in the Inferotemporal Cortex of the Macaque," J. Neurophysiol. 35, 1972, 96-111.

12) E.L. Schwartz, R. Desimone, T.D. Albright and C.G. Gross, "Shape Recognition and Inferior Temporal Neurons," Proc. Natl. Acad. Sci. (US) 80(18), 1983, 5776-5778.

13) K.S. Lashley, "In Search of the Engram," in Symposia of the Society for Experimental Biology, No. 4, Physiological Mechanisms in Animal Behavior, Academic Press, New York, 1950, 454-482.

14) R.M. Restak, The Brain: The Last Frontier, Bantam, New York, 1984, p. 54.

Computer simulation of Freud's counterwill theory: Integrating artificial intelligence and control systems approaches

Martin W. Denker, M.D., Karl E. Achenbach, Ph.D., and Donald M. Keller, Ph.D.
Department of Psychiatry and Behavioral Medicine
University of South Florida College of Medicine
Tampa, Florida 33612

ABSTRACT

Although there has been notable progress in Artificial Intelligence research, one area discussed in the early days of this field still remains relatively less well developed. Among the first to pose the problem were Simon (1967) and Arbib and Kahn (1969). The problem concerns the mechanisms of interaction between emotional regulatory systems and information processing operations. As a step toward developing such a simulation model, the authors extended an earlier model (Wegman, 1977) to describe the dynamics of interaction between individual thought processes, individual physiological and emotional processes, and key interpersonal relationships which influence individual thinking. This model was successful in describing several types of cognitive and emotional steady states in the dyadic interpersonal relationship (Denker, Achenbach & Keller, submitted for publication). The present work begins to test whether or not these apparently stable dyadic relationships accurately simulate the properties of a homeostatic system by means of studying the impact of perturbation and recovery on a selected sample of steady state dyadic systems. When such a control system is successfully modeled, then a hybrid simulation language will be used to describe both the contents of the information being processed and the regulatory factors which determine the rate of its processing. The hybrid model would then permit more accurate description of human behavioral sequences, intensities and rates.

INTRODUCTION

The work presented here describes the extended preparation of a continuous system simulation model which is intended primarily to be part of a larger, hybrid simulation model of human emotional and cognitive self-regulation, or personality system functioning. The work began in 1977 when it was proposed, using continuous system simulation methods (Forrester, 1968), that Freud's Counterwill Theory might not to be a simple model of one discrete type of behavior, but actually a large, complex dynamic model of all of human personality functioning, including human cognition and its interaction with motivation, behavior, and, to some extent, outside influences (Wegman, 1977). In order to continue and extend the translation of Freud's theories of human thinking, it was decided to attempt the stepwise construction of a large hybrid (discrete/continuous system) simulation model. In such a model, the physiological control system would be modeled by a continuous system submodel, and the cognitive system would be modeled by a discrete system submodel using Artificial Intelligence and information processing concepts.

The present investigators carried out the first step in this process by modifying the continuous system model of individual personality functioning developed by Wegman in 1977. Wegman's model seemed appropriate for the control system submodel because its behavior was consistent with clinical descriptions of a wide range of discretely different levels of human functioning. For example, under certain initial conditions, Wegman's model appeared to reach an equilibrium-like state which would appear to describe normal functioning. Under other initial conditions the control system either oscillated, had periodic collapses and recoveries, or shifted downward rapidly into a second equilibrium-like state at a lower level of functioning. The periodically collapsing and recovering state appeared to be a reasonable translation of the concept of a "nervous breakdown". The present investigators extended this model to describe the simultaneous behavior of the control systems of two individuals linked in a close personal relationship.

However, before undertaking the translation of the extended Wegman model into a new simulation language, it appeared important to conduct one further test of the general behavior and plausibility of this extended model. Accordingly, a form of perturbation upon both individual and dyadic models was decided upon. This perturbation was to mimic the impact of psychological stress on the individual both within and without the context of a supporting close personal relationship. If the extended control system did resist perturbation and/or did return to the pre-stress status, then homeostatic behavior could be said to be part of the control system submodel. This finding would support the decision to proceed with the translation.

Artificial Intelligence research approaches to these same types of human functioning seem clearly to offer promise. However, most Artificial Intelligence investigation of human functioning begins with the study of cognition, assuming relatively complete, and uninterrupted, rationality on the part of the simulated individual. And yet, most of these same investigations are quick to comment that the motivational and emotional factors which influence cognition should be taken into account for more advanced models (Simon, 1979).

Parallel with this conceptual problem has been a methodological problem. Artificial Intelligence research is usually conducted with discrete system simulation languages, while the limited study of emotional and motivational processes has used continuous system simulation languages. Recently simulation languages which provide the possibility of using both types of system simulation simultaneously, with provision for reciprocal influence between both submodels, has become more readily available. It thus seems timely to experiment with building an advanced Artificial Intelligence model of human cognition using the new hybrid technology.

METHODS

As mentioned, the starting point of the present investigation, with respect to both the theoretical model and the simulation methodology used, is the individual model of personality dynamics of Wegman (1977) extended to dyadic interpersonal relationships by Denker, Achenbach and Keller (submitted for publication). Concepts from certain parts of Freudian personality theory and from several different areas in social psychology were translated into equations using the System Dynamics method (Forrester, 1968).

The design of standard perturbation for the model was taken from the definition of psychosocial stress given by Caplan (1981). Caplan defined stress as a combination of increased emotional arousal and simultaneously decreased cognitive processing capacity. In order to model this process, the rate variable which determined emotional arousal level was increased by a factor of 50 percent and the rate variable which modeled the cognitive process of suppressing negative ideas was decreased by a factor of 30 percent. This model of stress was implemented in the form of a step function in which both of these factors were incremented or decremented simultaneously and instantaneously, held at the new values for a period of 40 time-units, and then instantaneously returned to their standard values. The standard time-unit has not yet been defined in real terms. However, the time horizon of the overall simulation project has been one designed to predict cycles of behavior extending over periods of weeks and months, so that one arbitrary time unit is probably on the order of magnitude of one day. Therefore, the time period simulated using this model of psychosocial stress is approximately six weeks.

Initially, the stress was applied to each of the examples of individual functioning originally demonstrated by Wegman, and each of the examples of dyadic equilibrium demonstrated by the present investigators after the extension of the original theory. After the responses were collected and examined, additional dyads were selected for study under the perturbation conditions. For any dyad being studied, the stress protocol called for testing each member of the dyad in isolation, then testing each member separately within the dyadic context, and then testing both members simultaneously in the dyadic context.

At this stage of the theory construction project, the goal was not to make specific predictions about which kind of dyad would respond in which way to the standard stress. The goal was a more general one, namely, to see if this approach would produce a wide variety of different, yet plausible, responses under different conditions. In particular, responses simulating homeostatic behavior by the system would be of greatest interest. These criteria of richness and plausibility have been used consistently by the present investigators as a method of determining whether or not to continue through each of the many steps involved in the construction of this large and complex model.

RESULTS

A wide variety of interesting and plausible reactions to stress was observed. Space considerations permit illustrating only a few.

Homeostatic behavior, the most interesting of the many stress responses, was seen in at least three different varieties. Figures 1A and 1B show both members of a dyad being stressed and undergoing a steady decline in function during the stress period (t=80 to t=120), with a "delayed" collapse 20 time-units after the stress is removed, followed by quick and complete recovery 40 time-units after the end of the perturbation. Stress to either individual Q or individual Z alone (not illustrated) produces no system response during the stress period ("resistance") followed by complete collapse when the stress is removed (model does not compute). In terms of clinical plausibility, this dyad was formed by two "unstable" individuals (periodically collapsing mode) who achieved a higher level of stability and stress resistance in their dyad than either individual could have expected to achieve in isolation. Also, each individual gained increased stress resistance in the dyadic context, but only if the stress is evenly distributed. If the stress is unevenly distributed, then Q and Z experience increased vulnerability.

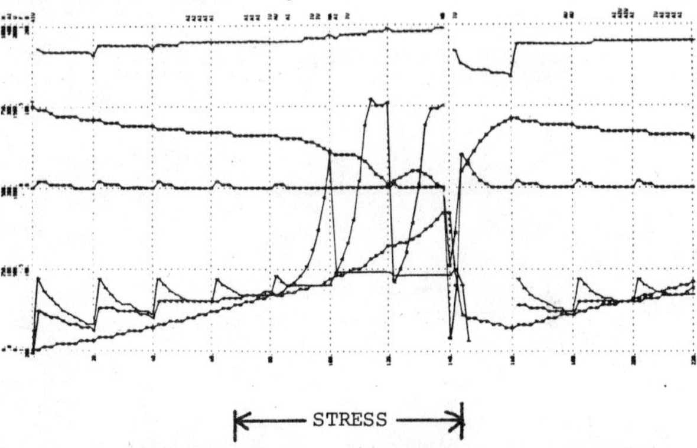

Figure 1A. Homeostasis in Q as return to equilibrium.

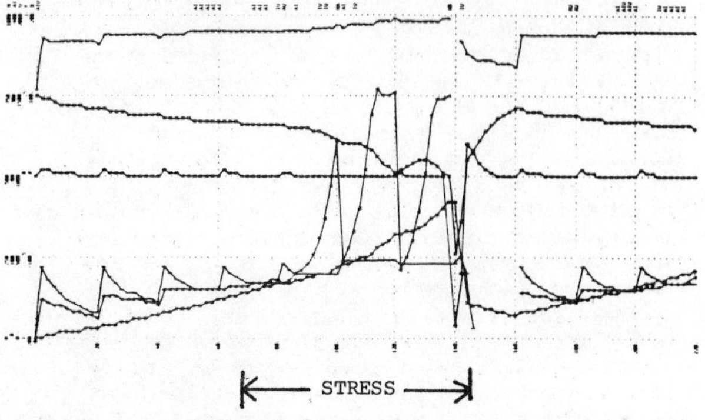

Figure 1B. Homeostasis in Z as return to equilibrium.

A second type of homeostatic response is seen in Figures 2A and 2B. Both members of a dyad again receive the same stress as in Figure 1, but the dyadic system reveals no response at all to perturbation. There is also no major response to stress on either individual Q or Z alone (not illustrated), although slight temporary stress-related changes (symptoms) are seen in individual Q, as in Figure 2A. Although the stability of this dyad is achieved at the cost of a low level function in Q, the resistance to stress is maintained no matter how the stress is distributed.

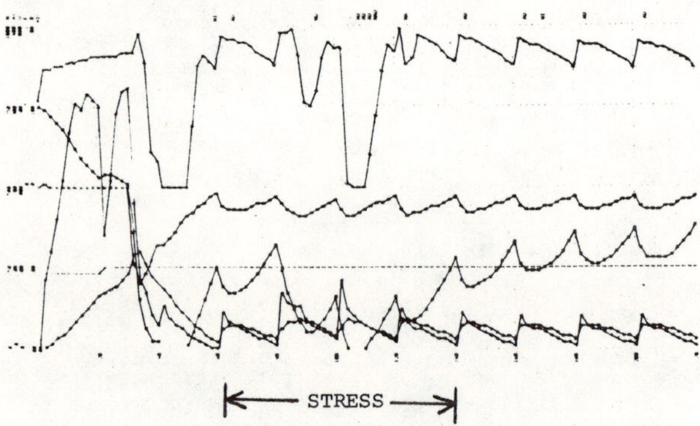

Figure 2A. Homeostasis in Q as resistance to departure from equilibrium.

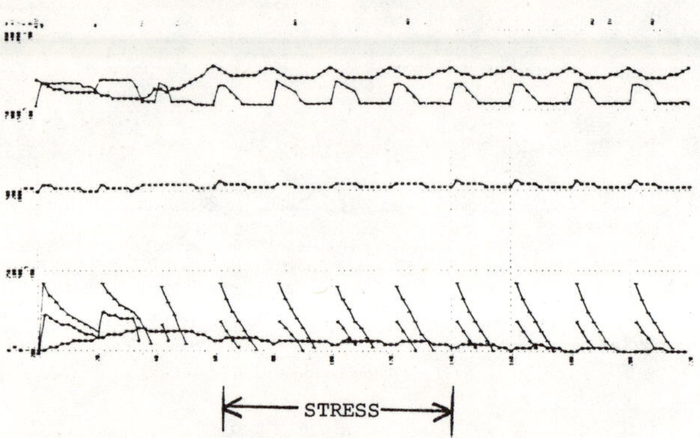

Figure 2B. Homeostasis in Z as resistance to departure from equilibrium.

A third type of homeostatic system response is seen in Figures 3A and 3B. Again both members of a dyad are undergoing the same stress. Individual Q, who functions slightly better in this dyad than when alone, and who would be expected to have mini-collapses in either condition at t=80, has an unexpected (paradoxical) major collapse at the removal of the stress (t=120), while individual Z shows no responses. The term "paradoxical" is used here because system function worsens, temporarily, after the stress is removed, at a time when function would have been expected to improve. Individual Q has no response when only individual Z is stressed (not illustrated). Individual Q has a more pronounced and delayed collapse when only Q is stressed in the dyadic condition (increased vulnerability, not illustrated). These types of paradoxical and/or delayed response appear to be related to the accumulation processes within the system. Individual Z shows no response to stress either in the dyadic or the isolated condition. Thus this relationship illustrates a complex partial form of "altruism". Both individuals function at a higher levels in the dyad than in isolation, although Q is still slightly unstable. When in the dyadic condition, Z holds onto these gains, even under stress, while Q suffers under stress, but not as much as in isolation.

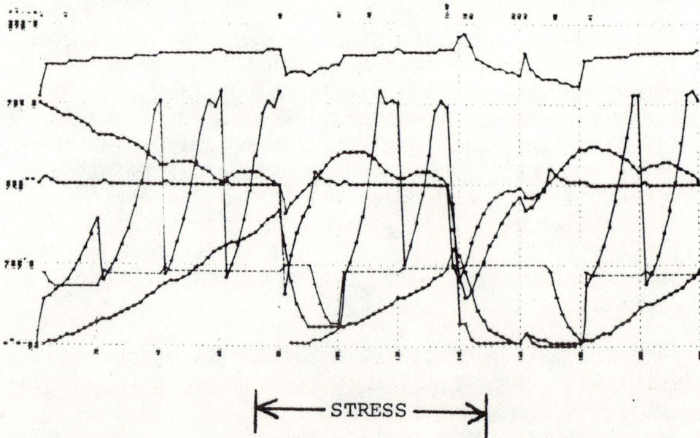

Figure 3A. Homeostasis in Q as resistance followed by delayed reaction.

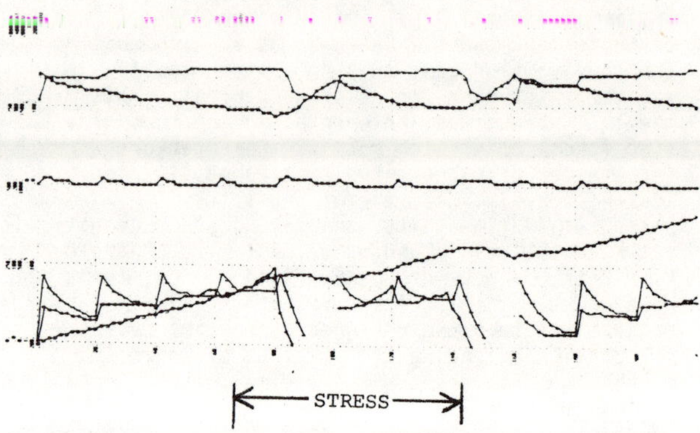

Figure 3B. Homeostasis in Z as resistance followed by delayed reaction.

Thus the results show a rich variety of homeostatic behavior, including recovery, vulnerability, resistance, shared resistance, delayed reactions and paradoxical reactions to stress. The results also show bilateral and unilateral types of responses to shared stresses. Such results continue to confirm the internal consistency of the modeling process originated by Wegman in 1977, and the extensions to date. The results continue to be clinically plausible (Caplan, 1981), and no large inconsistencies have yet emerged which will require major changes in the model.

DISCUSSION

Overall, the results appear to support the continuation of steps toward the hybrid simulation model of human cognition and personality functioning described above. In particular, the results suggest that the control system submodel of the proposed larger model seems to be a plausible control system with respect to its individual properties, its properties when linked to a second control system

in an interpersonal dyad, and its responses to
stress in the individual and the dyadic conditions.

The hybrid simulation language under consideration at the present time is SIMAN. It is planned
that the control system will be written into SIMAN
as the continuous system submodel, and the information processing part of the cognitive process will
be written in as the discrete system submodel. The
model building process will require the recoding of
the DYNAMO equations into FORTRAN within subroutine
STATE as the continuous component within the SIMAN
system model. A block diagram or flowgraph of the
information processing system will then be developed
by coding into the SIMAN language concepts presented
in the information processing models of cognition
developed by Simon (1979) and Malone (1975). It is
anticipated that certain advanced properties of the
SIMAN language, such as the modeling of transport
processes and the modeling of distances between
processing stations, may be used in modeling this
cognitive processing operation and may permit more
sophisticated models than have been previously
possible in languages other than LISP.

The next step, and possibly the most crucial
one theoretically, will be to combine the two
submodels. In part this will be done by writing
equations describing the changes imposed on the
continuous submodel from the discrete submodel
by means of assigning changes in switch functions,
tables and constants through the ASSIGN block
function. At the same time changes will be made
to the discrete component from the continuous
submodel by means of threshhold crossings which
trigger state events. These equations would then
constitute an attempt to model two simultaneous
and reciprocal phenomena of critical importance
in personality theory. First, the effect of
information processing on the regulatory system
will model how the individual learns new ways
to self-regulate. Simultaneously, the effect of
the regulatory system upon information processing
will model the phenomenon known as state-dependent
learning.

In addition, as often happens, simulation
methodology helps clarify theoretical and defini-
tional problems in the underlying discipline to
which the methodology is being applied. For
example, this model serves to separate out the time
course of the stress, response, and recovery phases
in each individual in the relationship (Mullan,
1981), as well as to separate out and visualize
the difficult case when the stressful event to
individual Q is the occurrence of a stressful
situation within individual Z who is, at the same
time, part of the social support system or stress
buffering mechanism for individual Q. So far, in
empirical research on stress and social support
mechanisms, it has proved difficult to define

precisely what stress is, to separate out direct
and indirect effects of stressors and social support
relationships, and to display the time courses of
the stress and adaptation reactions, particularly
when they may not be identical in timing with each
individual in a relationship. Simulation of these
processes may help to improve the design of future
research by first drawing a hypothetical "picture"
of what findings are to be searched for in the
research effort.

REFERENCES

Arbib, M.A., and Kahn, R.M. A developmental model
of information processing in the child.
Perspectives in Biology and Medicine,
1969, 12, 397-415.

Caplan, G. Mastery of stress: Psychosocial
aspects. American Journal of Psychiatry,
1981, v. 138, pp. 413-420.

Denker, M.W., Achenbach, K.E., and Keller, D.M.
Computer Simulation of Freud's Counterwill
Theory: II. Extension to elementary social
behavior. (Submitted for publication.)

Forrester, J.W. Principles of systems.
Cambridge, Mass.: Wright-Allen, 1968.

Malone, T.W. Computer simulation of two person
interactions. Behavioral Science, 1975,
20, 260-267.

Mullan, J.T. The (mis)meaning of life events in
family stress theory and research. Paper
presented at the Theory Construction and
Research Methodology Workshop, Annual Meeting
of the National Council on Family Relations,
St. Paul, Minnesota, 1983.

Simon, H.A. Motivational and emotional controls
of cognition. Psychological Review, 1967,
74, 29-39.

Simon, H.A. Information processing models of
cognition. Annual Review of Psychology,
1979, v. 30, 363-396.

Wegman, C. A computer simulation model of Freud's
counterwill theory. Behavioral Science,
1977, 22, 218-233.

ACKNOWLEDGEMENTS

The authors wish to express their gratitude to
the Department of Psychiatry, College of Medicine,
University of South Florida for support of this
research. Special thanks are due to Departmental
Chairman, Anthony J. Reading, and to Departmental
Research Assistant, Kathleen Vilmure.

Testing an expert system for manufacturing

Adel S. Elmaghraby, Richard S. Demeo and John Berry
Engineering Mathematics and Computer Science
Speed Scientific School
University of Louisville, Louisville, KY 40292

ABSTRACT

An expert system for flexible manufacturing has been developed as well as an interactive simulation software for such systems. This study demonstrates the use of simulation in evaluating the expert system.

Knowledge representation in this expert system is in the form of rules. Several sets of rules are used--one set of rules is included for optimizing critical path assignments and a second set is included for optimal floor layout. An interactive simulation software for manufacturing systems has been developed and interfaced to the expert system through a shared data base.

Integration of the simulation software with the expert system is discussed. Case studies showing bottlenecks and cost analysis of the manufacturing system are included. Results of these studies are analyzed to demonstrate the effects of various rules in the expert systems.

Essentially, simulation of the manufacturing system is used as a design aid for evaluating an expert system. This allows better understanding of the overall effect of various rules and gives directions for modifications and expansion of the knowledge base.

INTRODUCTION

The area of manufacturing has become very important. Increased productivity and flexibility are among the most important objectives with other industrialized nations in automating its factories. Directions in automation include the following:

1. Automatic machine tools, assembly machines, and material handling systems.

2. Computer control.

3. Data collection and Decision Support systems.

The research presented in this paper demonstrates an approach for implementing a decision support system for manufacturing. The methodology used is that of knowledge-based expert systems. Validity of the acquired knowledge is tested using traditional simulation techniques.

Goals of the expert support system are targeted to a plant manager. This expert system will assist first in deciding on the feasibility of a particular production, given the available resources. Second, on the assignment of resources to specific tasks. Finally, the expert system will provide optimal floor layout.

The simulation software is used to test results of a particular design for the manufacturing process. The user is allowed to create variations of the system through graphic interactions. Simulation and display of up to ten configurations for comparison is allowable.

SYSTEM DESIGN

The design philosophy of this system stems from its purpose as a decision support system, and from its development in a research environment. A decision support system requires a good user interface as well as the ability to store data and retrieve it. The use of a knowledge-based approach sets requirements of explicit knowledge representation and need for knowledge acquisition. The development of the system in a research environment has set more rigorous testing requirements.

Development of the overall system has resulted from a coordinated research effort. Design and development of a computer aided design tool [1,2] provided the basis for the expert system. The expert system was tested separately to demonstrate its sensitivity to its knowledge-base. A user interactive simulation of a flexible manufacturing model was developed and reported [3,4] and demonstrates the traditional use of simulation for evaluation. The integration of both systems required use of a shared relational database. And our results will demonstrate the use of simulation for system testing and evaluation of the knowledge-base of the expert system.

The Expert System

The expert system is based on knowledge representation in the form of rules. Several forms of rule definition have been utilized. Some rules are represented in the form of a matrix of numerical priority values, others are defined by a sequence of procedural actions. User control on some actions is possible in some cases on line through interaction, in other cases the rules can be changed by changing a rule-file. One common design requirement is the explicit representation of these rules. Two major reasons are typically cited for explicit representation:

a) for expert systems - this provides a flexibility and expandability of the system,

b) for support systems - this clears the system from accusations of hidden biases.

The expert system in this implementation is included in a package with graphic support and a refined user interface. The user of this system needs to define the available resources as well as the tasks to be performed for a particular job. Data can be saved for resources and tasks and retrieved when needed. Resource definition includes their capabilities, running cost, physical location on the factory layout as well as their reliability. Flexible resources with a limit of four different tasks are allowed. Task definition requires definition of possible resources to perform it together with information on its sequence in the overall job. Tasks allowed include assembly from multiple preceding tasks or splitting such as following metal cutting.

Three sets of rules were used in the implementation of the expert system. These rules are used for:

a) Rule-based task reassignment - a set of priority rules are included which would instantiate procedures until a solution is reached. These are mainly concerned with assigning the least flexible resources as well as the most abundant resources to perform tasks before considering more expensive or least available resources.

b) Rule-based task reassignment - a set of rules instantiated on demand to reassign resources. This process uses a critical path algorithm with rules for assigning bottleneck tasks to multiple resources if the production objective is time; otherwise, minimum resources for a job are kept.

c) Rule-based floor assignment - rules for floor layout are executed. These rules allow the inclusion of fixed resources on the factory floor and redesign the floor to minimize transportation among production cells. An alternate mode is included which assumes an unrestricted optimal floor layout.

The Simulation Software

The simulation software utilizes a database generated by the expert system to define an initial job design. The user interface is achieved through a menu system with color graphics and allows editing and loading previously defined jobs. The user is allowed to request up to ten different simulation runs with or without on line graphic display of activities. Automatic gathering of statistics is included and various job reports can be requested through the menus including comparative reports. Ideally a Monte Carlo simulation should be produced for each run to produce statistically significant results. Considering the need for graphics and the time needed for multiple runs, it was decided not to include it. A simple code modification can allow multiple runs. However, the user can choose to run ten simulations of the same job with the current system.

<u>RESULTS</u>

The main purpose of using this software is the comparison of various designs for job production. Testing performance of optimal designs of the expert systems were compared to those obtained by a human expert. Comparisons show that although the expert system performance is near optimal and is sensitive to changes in the knowledge-base, it can be improved. Knowing that the knowledge acquisition phase for these rules from experts has been preliminary, one can easily understand these

results. Actually, some of the cases were purposely generated to outperform the expert system.

Figure 1 represents layout and resource assignments using the expert system with simulation results. Simulations 1 and 2 are based on the system in Figure 1 and same results are given in Figures 2 and 3 to demonstrate the statistical dependence of the simulation giving different bottlenecks but not a very different average production time. Note that if the simulation was run for a longer time period, these differences will be irrelevant.

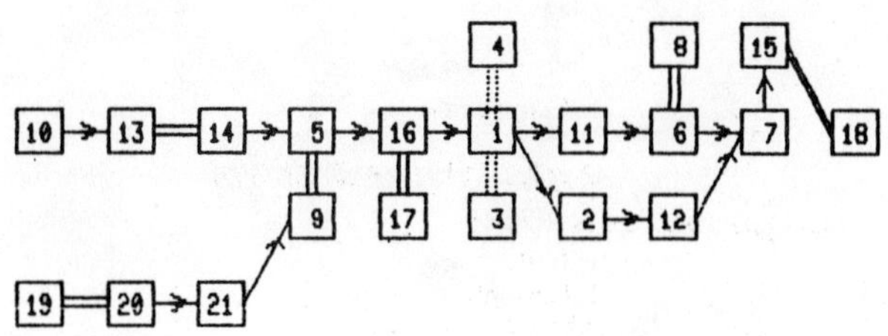

Figure 1

SIMULATION NUMBER 1

Total run time is:
    1 hours
    12 minutes
    50 seconds.
Total cost is $ 581.33
    100 parts where produced.
Average production time(sec.)   43.70
The three largest bottlenecks are:
  1. #21 - GRINDER , 94 parts/time 822
  2. #14 - ROBOT-S , 43 parts/time 585
  3. #13 - ROBOT-S , 38 parts/time 590

Figure 2

SIMULATION NUMBER 2

Total run time is:
    1 hours
    7 minutes
    11 seconds.
Total cost is $ 608.01
    100 parts where produced.
Average production time(sec.)   40.31
The three largest bottlenecks are:
  1. #21 - GRINDER , 86 parts/time 554
  2. # 5 - ROBOT-W , 29 parts/time 1814
  3. #14 - ROBOT-S , 28 parts/time 979

Figure 3

Figures 4 and 5 depict variations on the system generated by the expert system. In Figure 4a, resource #18 is removed thus performing its task by resource #15. Similarly in Figure 5a, resource #8 is also removed. Simulation results of these job designs are also presented in figures 4b and 5b respectively showing bottlenecks, cost and production times.

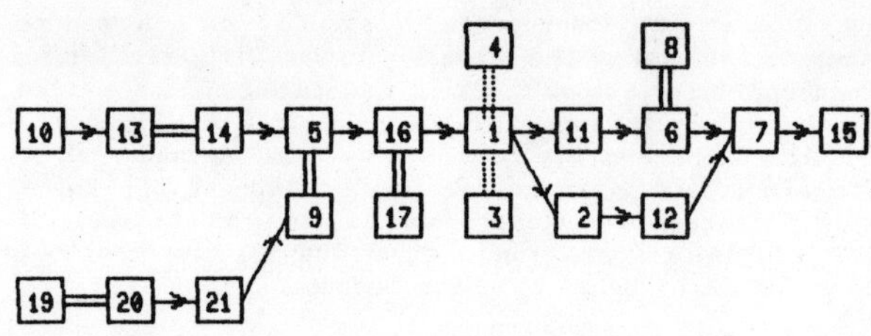

Figure 4a

```
    SIMULATION NUMBER  3

Total run time is:
      1 hours
      7 minutes
      28 seconds.
Total cost is $     559.00
  100 parts where produced.
Average production time(sec.)   40.48
The three largest bottlenecks are:
   1. #21 - GRINDER , 95 parts/time   371
   2. #14 - ROBOT-S , 35 parts/time   681
   3. #13 - ROBOT-S , 31 parts/time   843
```

Figure 4b

Figure 5a

```
    SIMULATION NUMBER  4

Total run time is:
      1 hours
      4 minutes
      59 seconds.
Total cost is $     535.15
  100 parts where produced.
Average production time(sec.)   38.99
The three largest bottlenecks are:
   1. #21 - GRINDER , 96 parts/time   331
   2. # 5 - ROBOT-W , 33 parts/time  1567
   3. # 9 - ROBOT-W , 32 parts/time  1541
```

Figure 5b

CONCLUSIONS

Analysis of these and other results suggests that simulation results can provide a learning tool. Learning can assist in redefining the rules for the expert system and finally assisting optimizing the manufacturing process. Current results suggest that a particular rule for assigning the same job to more than one resource to speed up production and eliminate bottlenecks need to be revised. Although improvement is needed, with the current set of rules, the system performs providing feasible, near optimal solutions.

Future plans for this project include cooperation with manufacturing experts. This is needed to improve the rule-base and provide more realistic test cases corresponding to actual manufacturing jobs.

ACKNOWLEDGEMENTS

Most of this work was performed for thesis requirements and was partially supported by grants from the University of Louisville Graduate School.

John Berry is currently with IBM at Rochester, Minnesota, and Richard Demeo is currently with the Naval Academy at Annapolis, Maryland.

REFERENCES

[1] John Berry and Adel S. Elmaghraby, "Computer aided Optimal Design Of A Flexible Manufacturing System", Proceedings of IEEE Southeastcon'84.

[2] John Berry, "Computer aided Optimal Design Of A Flexible Manufacturing System", Master of Science Thesis, University of Louisville, Louisville, KY, 1984.

[3] Richard Demeo and Adel S. Elmaghraby, "A User Interactive Simulation Of A Flexible Manufacturing System", Proceedings of IEEE Southeastcon'84.

[4] Richard Demeo, "A User Interactive Simulation Of A Flexible Manufacturing System", Master of Science Thesis, University of Louisville, Louisville, KY, 1984.

Human memory: A better understanding

RAYMOND P. O'CONNOR
BOEING COMPUTER SERVICES

Abstract

Humans come in many different sizes, shapes and colors. It is unreasonable to assume that there is one and only one way which human memory stores and retrieves data. The analogy to computer data storage provides a starting point for proposing human techniques.

A common computer data storage technique is hierarchical structured. The data is first classified into the appropriate structure and then stored at the proper level within the hierarchy. In 1960 David Ausubel proposed a technique for facilitating learning based on the assumption that human memory is hierarchical. If the subject is presented with statements containing an orderly arrangement of the generalized concepts called "Advanced Organizers" the subsequent learning would be facilitated. Over the last twenty-four years there has been a great deal of research to test Ausubel's theory. Of those studies which have been published, the results are mixed. Substantially less than half of the studies produced statistically signficant positive results. Frequently, negative results are measured. One possible conclusion to the lack of empirical proof of Ausubel's theory is that not all people have developed hierarchical memory storage techniques.

Computer data can be either sequential or direct. Studies of amnesia victims suggests loss of memory occurs for time slices. A patient may lose the ability to retrieve an event, a day or several weeks. Is it the actual data that is lost or the index (pointers)? Is amnesia cured when new indexes are established? The research offers no answer.

Computer data can be stored directly using an algorithm to compute pointers. Research indicates a correlation between the phonetic description of data and its retrieval. "Sounds like or starts with the letter" are memory retrieval keys. If the data was stored using phonetic keys its retrieval is facilitated by those keys. Research into the use of keys for retrieval has been generally favorably but not conclusive.

Computer data can be stored using a relative technique. The index is computed based on the relationship between two data items. The use of association to facilitate memory retrieval has been effectively used since the early Greeks. New data is associated with data whose index is well established. The association data becomes the key for the new data. Memory training courses teach association techniques with usually successful results.

If there are many techniques or strategies for the storage of data in human memory, how can individual's personal technique be identified? The Selchow & Righter Company game "Trivial Pursuit" offers an interesting and enjoyable technique for simulating individual memory search. The study of subjects during a "Trivial Pursuit" game indicates memory retrieval is most often an unconscious activity (an automatic process or an acquired habit?). Observation of subjects indicates:

1. Some searches take longer than others. (hierarchical structure)
2. They know the answer but they can not retrieve it. (indexed)
3. They answer correctly, immediately, but are unsure of it. (direct)
4. They list other individuals or facts related to the desired data until an association facilitates retrieval. (relational)

The computer program presented in Attachment I will provide an interactive game which is intended to induce an intensive search of memory. The questionnaire which follows the game attempts to identify an individual's personal memory technique.

Why Study Human Memory?

Each year billions of dollars are spent on corporate training programs. Training represents a major cost to the American business community. The effectiveness of that training must be a serious business concern. A better understanding of human memory may be the key to significant improvements in learning and thus the effectiveness of training. This paper presents a review of one of the most popular proposals for improving the effectiveness of learning, David Ausubel's Advance Organizer (AO).

Ausubel (1960) proposed the learning situation be modified to correspond to the way human memory works. According to Ausubel's theory, human memory is hierarchical. To make learning more efficient, the instructor should provide the students with a structure, organization or classification scheme into which the material to be learned will fit. Not only will learning be facilitated but long term retention will be significantly improved. Learning is improved because the instructor provides, in advance, the memory structure which the students must develop for memory storage and retrieval. The Advance Organizer is a learning aid which allows new material to be easily incorporated into the student's memory system. In technical terms, the human memory system is organized in terms of highly inclusive concepts under which are subsumed less inclusive concepts and information data. Learning is accomplished by incorporating new material into a previously learned, general idea. The Advance Organizer is a nucleus around which new material is processed and stored (Zeitonm, 1983). The content and organization of the Advance Organizer acts as a priming mechanism (Peverly, 1981).

Review of Educational Research

Since David Ausubel proposed the use of Advance Organizers, in 1960, there have been hundreds of empirical studies to test his theory. At each meeting of the American Education Research Association over the last eight years, there have been at least two papers presenting a test of Advance Organizer theory. Ausubel's proposal has been widely tested and seems to be held in high regard by the education profession. The Advance Organizer, often called "guided discovery", provides a clear, systematic approach to the lecture format of teaching. The lecturer presents the main points and a general outline of the lecture with the expectation that the outline and/or subsequent restatement of those main points will facilitate learning and subsequent recall.

What Research Says About the Advance Organizer

In 1975 Baines and Clawson published a review of 32 AO studies, only 12 of which were significantly positive. A 1976 review of AO studies (Hartley and Davis) indicated most favored AO while a 1977 review (Baker) indicated most failed to support the AO hypothesis. In 1978 Kozlow reviewed 99 AO studies, in which only 22 were significantly positive and 29 were actually negative. Also published in 1978 was a review by Wiesendanger in which 11 of 21 studies indicated negative results. Wiesendanger concluded that the student's effort to learn the AO actually interfered with the intended learning experience. In 1979 Mayer analyzed 44 AO studies. Mayer concluded that the AO has a positive but small effect on learning and retention. It is surprising that a theory which seems so intuitively sound is not upheld by empirical research.

Review of 35 AO Studies by this author produced findings comparable to the aforementioned reviews of the literature. Only 3 studies indicated results where one of two test condition groups had a significant improvement in learning when compared to a control group. Twelve studies found favorable but non-significant results while 3 had 1 of 2 test conditions positive. While only 2 studies indicated negative results, 17 studies concluded that the Advance Organizer hypothesis was not supported.

Advance Organizers: Significantly Improved Learning

Salmon (1977) studied 200 college students using a research design in which there were four groups, an AO group, a pretest group, an AO and pre test group, and a control group. When technical material was presented the AO groups learning was significantly better than the control group. There was no significance for the pretest only group. When prose passages were the material to be learned none of the treatment groups upheld the hypothesis.

Swadener (1977) set up three groups; AO, guided self study and control. The AO group learning was significantly better than both of the other groups. A subsequent test ten weeks after the study indicated the AO group performance was still significantly better. Snonffer (1979) used a research design similar to Swadener and reached similar conclusions. The AO group did significantly better in learning a science lesson. There was a positive but not significant improvement in learning history for the AO group.

Advance Organizer: Improved Learning

Of the 12 AO studies which indicated improvement in learning when an Advance Organizer is used, Lawton (1978) is typical. Lawton studied 237 rural elementary students. Those presented with AO did better than those who were not. There is no indication of a test of significance. When the subjects of a study are subdivided into groups based upon ability the results of using AO are mixed. Kamil (1977) found that the AO helped good readers but not poor ones. Meyer (1978) discovered high performers used an AO strategy where average and poor performers do not. R. E. Mayer (1976, 1977, 1978) has performed several studies as to the effect AO has on the learning of computer programmers. The groups which receive the AO do better on some but not all test conditions. The control group not required to learn the AO, did better on straightforward problems. Wilhite (1982) studied 104 college students and concluded that the AO improved learning except for high ability students (contradicting the conclusions drawn by Kamil and Meyer). Bluestone (1980), Hall (1977), Scarborough (1982) and Speigel (1976) performed studies which concluded the AO improved learning. Wright (1973) set up an experiment with 60 three to five year olds. Wright attempted to teach AO methodology. The 60 children were divided into three groups of 20. One group was taught to use an AO strategy. The second group received a lesson specially designed to facilitate learning and the third group, the control, received no training. The AO group did best.

Advance Organizer: No Effect On Learning

Seventeen studies were reviewed in which there were little or no measurable differences between the AO groups and the other test conditions or control groups. Edwards (1976), the most energetic, studied 409 third grade, 529 fourth grade and 425 sixth grade students. Some of the experimental groups scored lower than the control groups. Overall the difference between groups could be explained by chance probability. Smith (1980) concluded the AO had little or no effect on the learning of scientific material (contradiction of Snonffer's conclusion). The conclusion drawn by Kuhara (1980) is that there is no evidence that the AO helped in the formation of the relative schemata. The 14 other studies offered nothing to add to Kuhara's conclusion (Armbruster, 1976; Farr, 1975; Holley, 1980; Jacobowitz, 1980; Mecks, 1976; Miller, 1976; Peverly, 1981; Rains, 1976; Sagria, 1977; Smith, 1980; Stanley, 1977; Twining, 1981; Weinstein, 1977; Witte, 1981). Why is there such a variance between what seems to be a logical theory and the empirical research?

The Previous Experience Factor

Three studies identify previous experience as influencing factors. Rhodes (1975) in a study of 80 high school students failed to uphold the AO hypothesis. Rhodes observed that the group with high previous experience did better than the other groups. Allington (1976) reached the unsupported conclusion that readiness for learning is dependent upon adequate previous experience. Lederer (1982) in a study of college students observed that previous experience of the subjects seemed to have an effect on performance.

The Human Memory System

If learning is developing a strategy for the storage process, then teaching is helping individuals form strategies which are meaningful to them (their previous experience). The measure of learning is the retrieval process. It is helpful to describe the human storage and retrieval process, by comparison to computer storage. From a storage standpoint, information is received, encoded and placed onto a storage medium (human memory).

Computer Storage

A comparison to computer technology, while useful to help intuitively to explain the process, should not be taken too literally. The computer which functions equal to or better than the human mind is still science fiction. Information may be stored by a computer as stream of information. The process of retrieving the information is time consuming and ineffective. An individual may spend days before finally remembering the name of "that song". A computer has an index structure. Information is placed in computer storage and an index is updated as to the location of that information. Recognition may be compared to locating the information in the index. Association occurs when the information is actually retrieved from memory in a manner similar to the computer direct access method. An algorithm is designed which takes a cue or retrieval key and translates it into a location where the information has been stored. Information can then be accessed and retrieved from the disk directly. Computer information may be stored using indexed sequential access method. At the end of the information is the address of the next sequential item. The human memory system forms associations where different pieces of information are linked. (Tulving, 1974).

The human memory seems to function in a relative manner in regard to time. Amnesia, as a result of physical trauma to the brain, often results in the loss of the ability to retrieve information stored in memory for a period of time preceding the trauma. Basic skills are not forgotten, language abilities are still present, but the patient is unable to retrieve from storage any information for a particular chronological period. That period may be hours or weeks before the trauma. It may be all information before the trauma. However, information stored after the trauma is retrievable and the processes or schema by which the information is stored are still available to the patient (Spear, 1976).

One of the primary services being provided today by large mainframe computers is large database applications. A common technique for storage and retrieval of information in a large database is hierarchical structuring. The database has some process or algorithm by which the level or hierarchical internal structure is defined. Information is stored and subsequently retrieved from the database based upon its level within the hierarchy and the coding scheme by which the association of data is obtained. Information with similar meanings or high association is stored within a multi-level structure based upon its appropriate association. Additional knowledge builds upon the existing information stored in an individual's memory system. These hierarchical structures may form the basis of knowledge and the basis of future cognitive development. Different structures are formed for different information contents. It is necessary, when retrieving information from a hierarchical structure, to identify the appropriate structure, move down to the appropriate level, and then extract the information. Psychologists have attempted to identify the extent of hierarchical structuring by the time delay necessary to retrieve information which is most probably stored in this fashion (Spear, 1976, Myers, 1976).

The technique for learning based upon the hierarchical nature of human storage requires that an individual take information to be learned and first organize it. The memory process is one of forming a structure, classifying new information as to its appropriate place in that structure, and then storing it.

Learning To Learn

Since learning strategies are not taught to the general public, the strategy an individual utilizes for storage and subsequent retrieval of information is a highly personal activity. An individual may form a strategy for organizing information that results in an exceptionally successful career in college. Another individual may not adopt a strategy of learning, expend considerably more energy and yet fail to complete the work necessary to graduate from college. There has been research whereby psychologists have attempted to teach learning strategies, but unfortunately, the results of these experiments have been unsuccessful. Because every individual has an experience which is different from all other individuals, no one individual can impose his or her memory structure or encoding process upon another. However, it is possible to make all learners aware of those techniques which will lead to successful learning. The technique of encoding information can be learned. Memory is highly personalized activity. However, the process for improving storage and retrieval of information in memory can be taught. That process, if adopted, should greatly increase the effectiveness of a learning endeavor and should result in a decrease in the variability among student populations. This concept is offered as a basis for future research.

The computer program in Attachment I is designed to simulate the human memory by creating a situation where an individual performs an intense memory search. The questionnaire at the end of the game may provide information as to an individual's personal memory technique. The game could be used in a training environment to help the instructor to design Advance Organizers which match the previous experience of the students. It is my hypothesis that such an AO would greatly improve the effectiveness of a learning experience.

Allington, R. L. (1976). Teaching, learning, and reading in the middlegrade content areas. (Report No. CS003284). Paper presented at the annual meeting of the Kean College Reading Conference, Union, New Jersey, October 16, 1975. (ERIC Document Reproduction Service No. ED140213)

Armbruster, B. B. (1976). Learning principles from prose: A cognitive approach based on schema theory. (Report No. CS003188). Technical Report No. 11, Illinois University, Urbana; Center for the Study of Reading. (ERIC Document Reproduction Service No. ED134934)

Ausubel, D. P. (1960). The use of advance organizers in learning and retention of meaningful material. Journal of Educational Psychology, 51, 267-272.

Bluestone, M. A.; Herst, S. (1980). The effects of an advance organizer, text structure and a preview of structure on the learning and retention of prose material. (Report No. TM800279). Paper presented at the Annual Meeting of the AERA, Boston, April 7-11, 1980. (ERIC Document Reproduction Service No. ED189116)

Edwards, B. A. (1976). The effect upon comprehension of the presence and type of cognitive organizer and of the syntactic structure of recall questions. (Report No. CS003418). Ph.d. Dissertation, University of South Florida. (ERIC Document Reproduction Service No. ED140231)

Estes, W. K. (1982). Models of learning, memory, and choice. New York: Praeyer

Farr B. P. (1975). The effects of thematic organizers on comprehension. (Report No. CS203237, ERIC Document Reproduction Service No. ED136264)

Hall C. K. (1977). The effects of graphic advance organizers and schematic cognitive mapping organizers upon the comprehension of ninth grade students. (Report No. CS003558). M.Ed. Thesis, Rutgers, The State University of New Jersey. (ERIC Document Reproduction Service No. ED141779)

Holley, C. D. (1980). Employing intact and embedded headings to facilitate long-term retention of text. (Report No. TM800286). Paper presented at the Annual Meeting of AERA, Boston, April 7-11, 1980. (ERIC Document Reproduction Service No. ED189121)

Jacobowitz, T. (1980). Teaching previewing techniques to high school students. (Report No. CS006493, ERIC Document Reproduction Service No. ED212982)

Kamil, M. L.; Hanson, R. H. (1977). Perceptual versus remantic information processing in semantic category decisions. (Report No. CS151723) Paper presented at the Annual Meeting of the National Reading Conference, New Orleans, GA, Dec. 1-3, 1977. (ERIC Document Reproduction Service No. ED151723)

Kintsch, W. (1970). "Models for free recall and recognition:, In. D. A. Norman ed, Models of human memory, New York: Academy Press

Kuhara, K.; Hatano G. (1980). The effects of advance organizer and anticipation activity on the learning of texts. (Report No. CS006895). Paper presented at the Annual Meeting of the AERA, Boston, April 7-11, 1980. (ERIC Document Reproduction Service No. ED222884)

Lawton, J. F.; Wanska, S. K. (1978). Transfer effects of different types of critical reading to gifted elementary students. (ERIC Document Reproduction Service No. ED226318)

Mayer, R. E. (1976). Some conditions of meaningful learning of computer programming: Advance organizers and subject control of frame sequencing. Journal of Educational Psychology, 68, 143-150.

Mayer, R. E. (1977). Effects of instructional organization and sequencing on productive learning. (Report No. SP010969). Paper presented at the Meeting of the AERA, New York, April 4-8, 1977. (ERIC Document Reproduction Service No. ED137313)

Mayer, R. E. (1978). Can advance organizers counter the effects of text organization? (Report No. CS004117). Paper presented at the Annual Meeting of the AERA, Toronto, Canada, March 27-31, 1978. (ERIC Document Reproduction Service No. ED154376)

Mayer, R. E. (1979). Can advance organizers influence meaningful learning. Review of Educational Research, 49, 371-383.

Mayer, R. E. (1979). Twenty years of research on advance organizers. (Report No. TN810590). California University, Santa Barbara, Department of Psychology. (ERIC Document Reproduction Service No. ED206691)

Mecks, J. W. (1977). An investigation into the effects of "imbedded aids". (Report No. CS003903). Paper presented at the Annual Meeting of the National Reading Conference, New Orleans, LA., Dec. 1-3, 1977. (ERIC Document Reproduction Service No. ED150544)

Meyer, B. J. F.; and Others (1978). Use of author's textual schema: Key for ninth graders comprehension. (Report No. CS004009). Paper presented at the Annual Meeting of the AERA, Toronto, Canada, March 27-31, 1978. (ERIC Document Reproduction Service No. ED151771)

Miller, W. L. (1976). A consolidation of advance organizers and student-developed pre-reading questions as a method to aid the retention of written information. (Report No. CS003426). A Ph.D. dissertation, University of Northern Colorado. (ERIC Document Reproduction Service No. ED140233)

Myers, J. L. (1976). Probability learning and sequence learning, W. K. Estes ed., Handbook of Cognition and Memory, Volume 3, New York: Lawrence Erlbaum Associates

Peverly, S. T. (1981). The effects of diagram-before text vs. diagram-after text in the processing of novel text information. (Report No. CS006507). M.S. thesis, Pennsylvania State University. (ERIC Document Reproduction Service No. ED212995)

Rains, M. J.; Meinke, D. L. (1976). Effects of cognitive style, group structure, instructions, and training sequence upon acquisition and transfer in concept attainment. (Report No. TM006913). Paper presented at the Annual Meeting of the AERA, San Francisco, Calif., April 19-23, 1976. (ERIC Document Reproduction Service No. ED150184)

Rhodes, F. J. (1975). The effects of advance organizers on selected cognitive styles and different cognitive structures in a programmed learning task of grammatical usage. (Report No. CS203210). Ph.D. dissertation, Pennsylvania State University. (ERIC Document Reproduction Service No. ED134998)

Sagria, S. D.; Di Vesta, F. J. (1977). Additive effects from interspersed adjunct questions in prose text. (Report No. CS003328). Paper presented at the Annual Meeting of AERA, New York, April 1977. (ERIC Document Reproduction Service No. ED137733)

Salmon, R.; and Others (1977). The advance organizer concept: Some methodological questions. (Report No. TM006911). Paper presented at the Annual Meeting of the AERA, New York, April 4-8, 1977. (ERIC Document Reproduction Service No. ED150183)

Scarborough, J. D. (1982). Comprehending technical discourse: The effects of text organization, visuals, and performance. Man Society Technology, Nov. 1982, p. 25-26.

Smith, D. (1976). Position of inserted questions and ability in learning from prose. (Report No. CS203491). Ph.D. dissertation, Indiana University. (ERIC Document Reproduction Service No. ED144050)

Smith, D. A.; and Others (1980). The effects of narrative analogies on the acquisition of passages describing scientific mechanisms. (Report No. CS205685). Paper presented at the Annual Meeting of the Western Psychological Association, Honolulu, HI. (ERIC Document Reproduction Service No. ED189605)

Snonffer, N. K.; Thestlethwaite, L. L. (1977). The effects of the structured overview and vocabulary pre-tracking upon comprehension levels of college freshman reading physical science and history material. (Report No. CS005290). Paper presented at the Annual Meeting of the National Reading Conference, San Antonio, TX, November 29 - December 1, 1979. (ERIC Document Reproduction Service No. ED182728)

Spear, N. E. (1976). Retrieval of memory: A psychobiolgical approach, W. K. Estes ed., Handbook of Cognition and Memory, Volume 4, New York: Lawrence Erlbaum Associates

Speigel, D. L. (1976). Investigation of the effects of training in the use of adjunct aids on comprehension when reading factual materials, with attention given to measuring attitudes. (Report No. CS003432). A Ph.D dissertation, University of Wisconsin - Madison. (ERIC Document Reproduction Service No. ED140235)

Staley, R. K.; Wolf, R. I. (1977). Adjunct questions in prose learning: The effects of response mode. (Report No. TM006836). Paper presented at the Annual Meeting of the New England Educational Research Association, Manchester, New Hampshire, May 4-7, 1977. (ERIC Document Reproduction Service No. ED148896)

Swadener, E. B.; Lawton, J. T. (1977). The effects of two types of advance organizers presentation on preschool children's classification, relations and transfer task performance. (Report No. PS009833). Wisconsin University, Madison, Research and Development Center for Cognitive Learning. (ERIC Document Reproduction Service No. ED152413)

Thelen, J. N. (1979). Role of pre-reading in content learning. (Report No. CS004952). Paper presented at the Annual Meeting of the West Virginia University Reading Center, Morgantown, WV, June 27-29, 1979. (ERIC Document Reproduction Service No. ED174962)

Thorndyke, P. W. (1981). Schema theory as a guide for educational research: White knight or white elephant? (Report No. SP019078). Paper presented at the Annual Meeting of the AERA, Los Angeles, April 1981. (ERIC Document Reproduction Service No. ED209227)

Tulving, E. (1974). Cue-dependent forgetting, American Scientists, 62, pp. 74-82.

Twining, J. E. (1981). Implications of schema theory for the guided reading of short stories. (Report No. CS006130). Paper presented at the Annual Meeting of the National Council of Teachers of English, Boston, Nov. 20-25, 1981. (ERIC Document Reproduction Service No. Ed311929)

Weinstein, C. E. (1977). Cognitive elaboration learning strategies. (Report No. TM006349). Paper presented at the Annual Meeting of the AERA, New York, April 4-8, 1977. (ERIC Document Reproduction Service No. ED144953)

Wiesendanger, K. D.; and Others (1979). A summary of studies related to the effects of question placement on reading comprehension. (Report No. CS005158). (ERIC Document Reproduction Service No. ED179934)

Wilhite, S. C. (1982). Pre-passage questions: The influence of structural importance. (Report No. CS006605). Center for the Study of Reading, Illinois University, Urbana. (ERIC Document Reproduction Service No. ED215311)

Witte, P. L. (1981). Glossing content-area texts: A vehicle for inservice training. (Report No. CS006379). Paper presented at the Annual Meeting of the American Reading Forum, Sarasota, FL, Dec. 10-12, 1981. (ERIC Document Reproduction Service No. ED210633)

Wright, J. C; Vlietstra, A. G. (1973). Attention and cognitive styles. (Report No. PS008778. (ERIC Document Reproduction Service No. ED129408)

Zeitonm, H. H. (1983). Advance organizer research: One step further. (Report No. SE040739). (ERIC Document Reproduction Service No. ED226996)

```
005 DIM A$(10,4),E$(10,4),S$(10,4),H$(10,4)
006 DIM G$(10,4),C$(10,16),X$(10),Y$(10)
010 ? 'WELCOME TO TOT.'
011 ? 'TOT (TIP OF THE TONGUE) DESCRIBES THAT'
012 ? 'SPECIAL SITUATION WHEN YOU KNOW YOU '
013 ? 'KNOW THE ANSWER BUT YOU JUST CAN NOT'
014 ? 'REMEMBER IT. DURING A TOT SITUATION'
015 ? 'YOU EXPERIENCE AN INTENSIFIED MEMORY'
016 ? 'SEARCH. PLAYING THE TOT GAME IS'
017 ? 'INTENDED TO SIMULATE A TOT STATE.'
018 Y$(1) = 'HIERARCHICAL'
019 Y$(2) = 'CHRONOLOGICAL'
020 Y$(3) = 'ASSOCIATION' Y$(4) = 'OPTICAL'
022 Y$(5) = 'PHONETIC'     Y$(6) = 'MUSICAL'
024 Y$(7) = 'QUEUED' Y$(8) = 'ENVIRONMENTAL'
026 Y$(9) = 'SMELL' Y$(10) = 'FREE ASSOCIATION'
110 E$(1,1)='CO-STAR OF MARILYN'
111 E$(1,2)=' MONROE IN THE'
112 E$(1,3)=' SEVEN YEAR' E$(1,4)='ITCH?'
114 C$(1,1)='A: BRUCE JENNER'
115 C$(1,2)='B: TOM EWELL'
116 C$(1,3)='C: GARY COOPER'
117 C$(1,4)='D: DEAN MARTIN'
118 REM YOU CAN ADD THE NEXT ( QUESTIONS
210 S$(1,1)='WHAT TEAM PLAYED'
211 S$(1,2)=' THE MOST
212 S$(1,3)=' WORLD SERIES '
213 S$(1,4)='GAMES?'
214 C$(1,5)='A: YANKEES' C$(1,6)='B: DODGERS'
216 C$(1,7)='C: RED SOCKS'
217 C$(1,8)='D: WHITE SOCKS'
220 REM YOU CAN ADD THE NEXT 9 QUESTIONS
410 H$(1,1)='THE NORTH CALLED '
411 H$(1,2)=' IT THE BATTLE OF '
412 H$(1,3)=' BULLRUN. WHAT DID'
413 H$(1,4)=' THE SOUTH CALL IT?'
414 C$(1,9)='A: GETTYSBURG'
415 C$(1,10)='B: MANASSAS'
416 C$(1,11)='C: ANTIGUM'
417 C$(1,12)='D: BULL RUN'
420 REM YOU CAN ADD THE NEXT 9 QUESTIONS
610 G$(1,1)='WHAT ARE THE' G$(1,2)=' CITIZENS '
612 G$(1,3)='OF NEW ZEALAND' G$(1,4)=' CALLED?'
614 C$(1,13)='A: YANKEES' C$(1,14)='B: BRITS'
616 C$(1,15)='C: KIWIS' C$(1,16)='D: AUSSIES'
619 REM YOU CAN ADD THE NEXT 9 QUESTIONS
700 T = 0 A$(1,1) = 'B' A$(1,2) = 'A'
960 A$(1,3) = 'B' LET A$(1,4) = 'C' ? ' '
01005 ? 'SELECT A NUMBER FROM 1 TO  10'
01010 INPUT S IF S<11 THEN 01100
01030 ? 'YOUR SELECTION IS TOO HIGH.'
01040 ? 'TRY AGAIN.' GOTO 01000  ? ' '
01110 ? 'TO SELECT A CATEGORY ENTER:'
01120 ? '1 FOR ENTERTAINMENT 2 FOR SPORTS '
01125 ? '3 FOR HISTORY, 4 FOR GEOGRAPHY.'
01130 INPUT C IF C=1 THEN 06100
01150 IF C=2 THEN 07100 IF C=3 THEN 08100
01170 IF C=4 THEN 09100
01180 ? 'YOUR SELECTION IS NOT A 1, 2, 3, OR 4.'
01190 GOTO 1100
06000 S = S + 1 C1 = 0
06020 IF S < 11 THEN 06100  S = S - 10
06100 ? E$(S,1);E$(S,2);E$(S,3);E$(S,4)
06140 ? 'ENTER ANY CHARACTER TO CONTINUE'
06145 ? 'ENTER END TO STOP.'
06150 INPUT Z$  IF Z$='END' THEN 90000
06160 ? 'ENTER A, B, C, OR D FOR THE ANSWER.'
06170 ? C$(S,1)  ? C$(S,2)
06180 ? C$(S,3)  ? C$(S,4) INPUT R$
06200 IF R$ = A$(S,C) THEN 06500
06210 IF C1 = 1 THEN 06000
06220 LET C1 = 1 ? 'TRY AGAIN.' GOTO 6160
06500 IF C1=1 THEN 06520  T = T + 5
06520 LET T = T + 5  GOTO 6000
06600 REM END OF ENTERTAINMENT SECTION

06900 REM THE 7000 8000 9000 SECTIONS ARE EXACT
06910 REM DUPLICATES OF SECTION 6000
90000 ? 'YOUR SCORE IS ',T
90005 ? ' '
90010 ? 'IF THE TOT GAME DID NOT SIMULATE A TOT '
90020 ? 'STATE TRY TO THINK OF A TIME WHEN YOU'
90030 ? 'WERE TRYING TO THINK OF SOMETHING WHICH'
90040 ? 'WAS RIGHT ON THE TIP OF YOUR TONGUE.'
90050 ? 'PLEASE ANSWER THE FOLLOWING QUESTIONS  '
90060 ? 'AS TO YOUR PERSONAL TECHNIQUE FOR'
90070 ? 'MEMORY SEARCH AND RECALL.'
90080 ? ' '
90085 ? 'AFTER EACH QUESTION ENTER A FOR ALWAYS,'
90090 ? 'B FOR FREQUENTLY, C FOR SOMETIMES'
90100 ? 'D FOR INFREQUENTLY AND E FOR NEVER.'
90105 ? ' '
90110 ? '1. HIERARCHICAL: YOU FOLLOW AN OUTLINE,'
90120 ? 'OR CLASSIFICATION STRUCTURE SUCH AS'
90130 ? 'LEADING MAN, RUSTIC,VIOLENT, GOOD-GUY.'
90140 INPUT X$(1)
90150 ? '2. CHRONOLOGICAL: YOU THINK OF OTHER'
90160 ? 'EVENTS WHICH OCCURRED AROUND THE SAME'
90170 ? 'TIME AS THE EVENT TO BE RECALLED.'
90180 INPUT X$(2)
90190 ? '3. ASSOCIATION: YOU THINK OF A COSTAR'
90200 ? 'OF THE PERSON WHOSE NAME YOU SEEK.'
90210 ? 'YOU THINK OF OTHER MOVIES IN WHICH'
90220 ? 'THEY STARED.'
90230 INPUT X$(3)
90240 ? '4. OPTICAL: YOU PICTURE A SCENE IN THE'
90250 ? 'MOVIE. THE ENVIRONMENT AROUND THE ACTOR'
90260 INPUT X$(4)
90270 ? '5. PHONETIC: YOU GO THRU THE ALPHABET'
90280 ? 'LETTER BY LETTER UNTIL YOU REACH THE'
90290 ? 'FIRST LETTER OF THE ANSWER.'
90300 INPUT X$(5)
90310 ? '6. MUSICAL: YOU THINK OF THE MUSICAL'
90320 ? 'ARRANGEMENT THE RECALL THE TITLE. THE'
90330 ? 'TITLE SONG HELPS YOU TO RECALL THE '
90340 ? 'DETAILS OF THE MOVIE.'
90350 INPUT X$(6)
90360 ? '7. QUEUED: WHEN YOU STORED THE DATA YOU'
90370 ? 'ASSOCIATED IT WITH A KEY PHRASE. RECALL'
90380 ? 'IS FACILITATED BY THINKING OF THE QUEUE'
90390 INPUT X$(7)
90400 ? '8. ENVIRONMENTAL: WHAT WAS THE '
90410 ? 'WEATHER OR ENVIRONMENT AT THE TIME OF'
90420 ? 'THE DESIRED EVENT?'
90430 INPUT X$(8)
90440 ? '9. SMELL: THE SCENT OF A PERFUME IS'
90450 ? 'ASSOCIATED WITH PERSONS OR EVENTS.'
90460 INPUT X$(9)
90470 ? '10. FREE ASSOCIATION: YOUR PERSONAL'
90480 ? 'MEMORY PROCESS SEEMS TO BE RANDOM'
90490 ? 'ALMOST AUTOMATIC (HABITUAL). WITHOUT'
90500 ? 'DIRECTION OR ORDER YOUR MEMORY IS '
90510 ? 'SEARCHED. INFORMATION FROM TOTALLY'
90520 ? 'DIFFERENT SOURCES/STRUCTURES SEEMS'
90530 ? 'TO BE EASILY ASSOCIATED.'
90540 INPUT X$(10)  ? ' '
90555 ? 'YOUR PERSONAL MEMORY STYLE IS'
90560 FOR X = 1 TO 10
90570 IF X$(X) = 'A' GOTO 90590
90580 GOTO 90600    ? Y$(X)
90600 NEXT X        ? ' '
90608 ? 'YOU ALSO USE TECHNIQUES WHICH ARE'
90610 FOR X = 1 TO 10
90615 IF X$(X) = 'B' THEN GOTO 90630
90620 GOTO 90650  ? Y$(X)
90650 NEXT X       ? ' '
90800 ? 'THANKS FOR PLAYING TOT'
99990 END
```

OPS5 as an electronic warfare design tool

Andrew Borden, Hugh Kao
Wayne Neyman, Edward Wojciechowski

ITT Avionics Division
390 Washington Avenue
Nutley, N.J. 07110

ABSTRACT

An Electronic Warfare expert system was developed for managing the defenses of an aircraft. OPS5 was selected as the expert system development language. Our efforts showed that OPS5 provided (1) a straightforward mechanism for translating the system designer's knowledge into the expert system's data base and (2) a built-in control structure for sequencing the flow of the system. These features, not available from conventional programing languages, allowed us to rapidly develop an aircraft defense system. Specifically, this system has the capability of detecting and recognizing multi-spectral threats, resolving sensor ambiguities and selecting appropriate countermeasures under system resource constraints.

1. EXPERT SYSTEMS AND THEIR APPLICATIONS

An expert system is a set of algorithms which efficiently apply a large body of knowledge to a complex data base. We will restrict our attention to "rule-based" systems, i.e. systems in which the knowledge is expressed as "if-then" statements. The system performs a pattern-matching analysis to determine which rules are applicable to the current configuration of the data base. The result of the rule application is to modify the data base in some way. A rule interpreter performs the pattern matching task and also applies a rule prioritization strategy to reduce the number of applications needed to reach a solution or goal state. The "expertness" of the system includes not only the knowledge embedded in the rules, but also the strategy which is based on "rules-of-thumb" or heuristic principles similar to those a human expert might use to limit a search space or resolve ambiguity. A common property of expert systems is the ability to explain their own behavior by tracing the application of rules and providing an audit trail through the changing configurations of the data base. In an expert chess-playing automaton for example, the data base is the set of reachable board configurations and the rulebase consists of the rules of chess. The rule interpreter determines which rules are applicable (which moves are currently legal) and applies an evaluation function to select the "best" one.

Obviously, if the data base is small, if the data is consistent and complete, and if the data does not vary except by application of rules, then no "expertness" is needed. Conversely, if the data base is very large (chess), if data is incomplete or inconsistent (medical diagnosis), if data elements change dynamically (process control), the application of a high level of expert knowledge is appropriate. Some observations about expert systems are the following:

- They do not modify their own rules
- They can grow incrementally by the addition of new rules
- Human intuition into rule interaction disappears at about 70-100 rules
- Existing expert systems vary from about 400-1000 rules
- An expert system cannot be constructed unless there exist human experts. An expert system may be faster and more reliable than the experts who created it, but it cannot be smarter

Expert systems have been successfully applied to the domains of medical diagnosis (MYCIN), mineral exploration (PROSPECTOR), mathematical analysis (MACSYMA), chemical analysis (DENDRAL) and computer system configuration (RI XCON).

2. THE OPS5 EXPERT SYSTEM DEVELOPMENT LANGUAGE

There exist a number of expert system design tools written in LISP. OPS5 is one such tool [1, 2]. It was developed at Carnegie-Mellon University and has been used to develop several successful system including R1-XCON. The data base WORKING MEMORY in OPS5 consists of sets of objects and attribute-value pairs. For example an object might be a radar emitter and an attribute-value pair might be (pulse repetition interval, 1000 microseconds). The rule base contains statement consisting of a left-hand-side (LHS), the "if" part; and a right-hand-side (RHS), the "then" part. The control structure consists of a "recognize-apply" procedure which attempts to match the LHS with a pattern in the current data base. When a match occurs, a rule is said to fire and the RHS is applied to the data base. The usual effect is to MAKE an (object, attribute-value) combination or to MODIFY an existing one to assign or change the value of an attribute. (MAKE and MODIFY are reserved words in OPS5). Examples of OPS5 applied to the aircraft defense problem will be given later.

The principal advantage of OPS5 is its simplicity. For example, rules can be added, deleted or changed without a need to modify the flow of the overall system. The OPS5 language takes care of the system flow. It is thus feasible to pursue a bottom-up design approach in which rules are added one-at-a-time, evaluated and kept or discarded as appropriate. OPS5 does record and report on the order in which rules fire so as to provide an audit trail of the decision making process.

The principal disadvantage of OPS5 is that the rule interpreter consists only of the "recognize-apply" cycle and two simple strategies to select one from among all the currently applicable rules (also called productions). Strategies to reduce the solution search space must be built into the rules or developed as a modification to OPS5 itself.

Another limitation of OPS5 is that it is difficult to use in the context of a real-time scenario driver. The impact of this shortcoming is that it becomes difficult to exercise the system under a variety of conditions and to evaluate time sensitive factors such as system response time. We have created a pseudo-time environment for our system and it will be described in Section 5.

It is possible to build an expert system which is more generic and more adaptable by placing certain types of "knowledge" in the data base instead of the rule base. If, for example, an expert system for a fighter aircraft has several versions pertinent to different missions or scenarios, mission specific data may be placed in a table in the data base. Generic productions in the rule base operate to match input data to the table. The generic productions would be changed infrequently and only by system designers. The table can be changed readily (within limits) by suitably trained, operationally-oriented personnel.

3. REQUIREMENTS FOR A MULTI-SPECTRAL AIRCRAFT DEFENSE SYSTEM

A modern tactical aircraft faces a dense, sophisticated, multi-spectral threat. It can be tracked by a radar or an infra-red, optical or electro-optical system on the ground or on another platform. The tracking system can initiate the firing of anti-aircraft artillery (AAA) or launch a missile. The missile can be guided by command from the tracking system or seek energy reflected by (or emanating from) the target aircraft. In order to survive, the penetrating tactical aircraft must employ an array of coordinated countermeasures to defeat one or more actions that the threat system must perform as part of its critical path.

The critical actions of the the threat system are usually called "key-states" and they are characterized by observable manifestations which can be used to trigger countermeasures of the right type at the appropriate time. Table I gives a listing of key states and associated observables for a hypothetical missile system.

In order to counter this missile system, a sensor suite is required to recognize key state transitions when they occur. The sensors might include RF (for radar detection and identification), infra-red (IR), laser detector, laser radar and electro-optics (TV). The sensors must cue one another as well as the countermeasures. For example, the RF sensor might direct a tracking IR sensor to search a given quadrant if there are indications that a missile launch is occurring.

Countermeasures depend on the type, spectral characteristics and phase of the engagement. Some countermeasures are used together to operate synergistically or to provide backup. For example, range deception against a radar makes it more vulnerable to angle tracking deception. Expendables (metallic strips or flares) may be used if on some deceptive measures are not successful.

Figure 1 shows the dependencies between the different sensors and countermeasures and their application against aspects of missile associated threat systems. The specific nature of the countermeasures and the parts of the spectrum in which they operate are not identified, but the general outline of the linkages is realistic.

Four aspects of the aircraft defense problem are appropriate for expert systems application.

First, many radars have parameter ranges in common and the attempt to identify them unambiguously can lead to large search spaces causing unacceptably long system response times.

Second, sensor data varies dynamically and is often incomplete and inconsistent.

Third, there are more situations than can ever be anticipated. Some may involve simultaneous engagement by several threats. The problem of selecting an optimal mix of countermeasures is one which challenges human experts, even given unlimited time to solve it. The selection of a sub-optimal countermeasures set in

Table I. Missile System Key States and Associated Observables

KEY STATE	FUNCTION	OBSERVABLES
ACQUISITION	DETERMINE TARGET LOCATION	RADAR SIGNAL WITH IDENTIFIABLE PARAMETERS
TRACKING	ESTABLISH ACCURATE TARGET LOCATION	CHANGE IN RADAR SCAN ASSOCIATED WITH TRAN-SITION TO TRACKING MODE
MISSILE LAUNCH	LAUNCH OF MISSILE WITHIN AN ENVELOPE THAT PROVIDES A HIGH PROBABILITY OF SUCCESSFUL INTERCEPT	APPEARANCE OF MISSILE GUIDANCE SIGNAL. INFRA-RED SIGNATURE OF MISSILE LAUNCH
MISSILE GUIDANCE OR SEEKING	GENERATING STEERING COMMANDS TO GUIDE THE MISSILE TO THE TARGET	MODULATION OF MISSILE GUIDANCE SIGNAL. INFRA-RED SIGNATURE OF MISSILE ENGINE

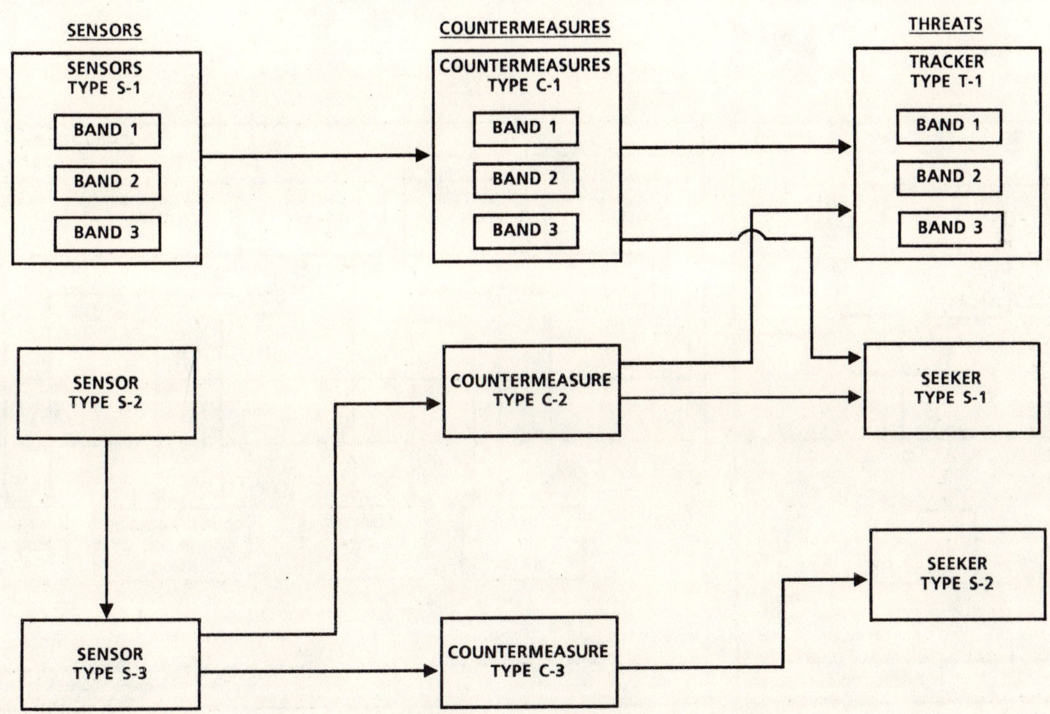

Figure 1. Sensor Cueing and Countermeasures Application

3-84-0213-01

real-time is a most suitable expert system task.

Finally, countermeasures compete for resources. When the resources (duty-cycle, power available, numbers of expendables) are exhausted, degraded modes or less-than-desired combinations must be used. This task can be made more difficult by component failures. The tasks of system self diagnosis, dynamic reconfiguring and optimal workarounds appear to be amenable to rule-based system attacks.

4. AN ELECTRONIC WARFARE EXPERT SYSTEM

An electronic Warfare (EW) expert system software program was developed for managing the defenses of an aircraft. This expert system has the capability to detect and recognize multispectral threat signals, resolve sensor ambiguities, and select appropriate countermeasures subject to system resource constraints. OPS5 was used as the expert system development language.

For rapid system development, OPS5 is an ideal tool. Expert knowledge can be entered into the data base in a straightforward manner, in the form of a flat list (i.e. no embedded structure). The control mechanism for selecting and implementing the rules is built-in to the language. This frees the system developer from worring about the detailed controls and sequencing of the production rules. Thus, in a system development phase, rules can be tried and tested for effectiveness without a need to modify the structure of the expert system.

The architecture of this EW expert system is divided into two parts: a working memory and a production memory. The working memory contains the information on emitter characteristics, countermeasure responses, sensor status, emitter identification status, and system resource status. The production memory is a collection of rules (also called productions) in the form of "IF-THEN" statements that operate on the working memory.

Figure 2 illustrates the structure of a computer simulation used to exercise the EW expert system. In the figure, a simulation driver, written in OPS5, time steps a system clock and selectively activates emitters in accordance with one of the scenarios in the Emitter Scenario data base. Scenario Update rules performs this activation process. The Scenario Update rules drive the expert system inputting data to the expert system's Sensor Status data base.

The expert system executes the Emitter ID rules to match the data in the Sensor Status data base with the data in the

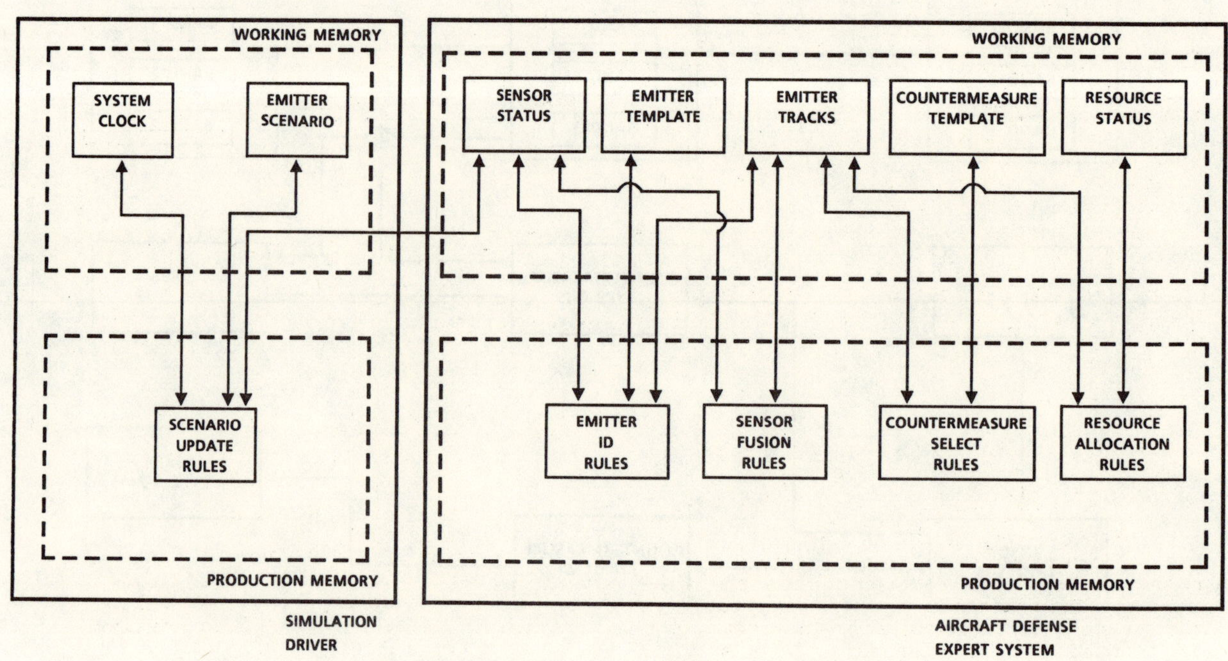

Figure 2. An Aircraft Defense Expert System Simulation

3-84-0213-02

74

Emitter Template data base. Sensor Fusion
rules govern the generation of emitter
track files located in the Emitter Track
data base. These track files contain all
correlated information from different
sensors that pertain to one threat system.

For those emitters with either
ambiguous identification or which require
additional sensor input, the Sensor Fusion
rules will either resolve the ambiguity
through multi-sensor correlation or cue
sensors to gather additional data. The
Countermeasure Select production rules
fire when emitter tracks are established
in the Emitter Tracks data base. It
matches a countermeasure technique to an
emitter track based on the data stored in
the Countermeasure Template data base.
The Resource Allocation rules match the
selected countermeasure against the
currently available resources needed to
generate the countermeasure responses. If
a conflict exists, either a different
countermeasure technique is selected or
the response to lower priority emitter is
dropped.

A sample output from the expert system
simulation is shown in Table II. The
scenario consisted of two weapon systems
each with radars acquiring and tracking
the penetrating aircraft. The expert
system on the aircraft detected the
emitters, established track files on the
emitters, identified them and generated
countermeasures (CM) against the emitters.

In the case of Emitter 1, the
countermeasure generated was not able to
prevent a missile from launching against
the aircraft. The staring sensor first
detected the missile launch. The expert
system then cued a scanning sensor to
establish an accurate track on the
missile's trajectory and calculated the
missile's range from the aircraft. An
accurate time-to-go (TGO) on the missile's
flight was then computed, followed by a
timely launch of expendables causing the
missile to miss the aircraft. Although
only two threats were simulated in this
scenario, the number of emitters that can
be considered is essentially unlimited.
The system can accept a realistic scenario
with a dense and complex multi-spectral
environment.

5. CONCLUSION

The utility of OPS5 as an EW expert
system development language was
convincingly demonstrated. It allowed the
system designers to rapidly build a
working expert system. As demonstrated in
the EW expert system for aircraft defense,
rules can be tried and tested for
effectiveness without a need to develop a
system control structure. Knowledge can
be encoded into the expert system's data
base in a straightforward fashion, in the
form of a flat list, without a need to
design complicated data base structures.
Our experience showed that it took only a
few weeks to develop a relatively
sophisticated EW expert system using OPS5
programming language. Whereas in the
conventional approach, considerably more
time is needed to develop a system with
the same sophistication.

Table II EW Expert System Output

```
START OF SIMULATION
: 1
: 2
EMITTER 1 ACTIVE
.....TRACK 1 WAS ESTABLISHED AND
      IDENTIFIED AS
        RADAR_1 AT TIME 2
.....RF CM APPLIED TO TRACK 1
: 3
: 4
: 5
EMITTER 2 ACTIVE
.....TRACK 2 WAS ESTABLISHED AND
      IDENTIFIED AS
        RADAR_2 AT TIME 5
.....RF CM APPLIED TO TRACK 2
: 6
: 7
: 8
MISSILE LAUNCH AT TIME 8
.....STARING SENSOR ACTIVATED
.....STARING CORRELATED TO TRACK 1 AT
      TIME 8
.....TGO UPDATED ON MISSILE_1
.....SCANNING SENSOR CUED AT TIME 8 FOR
      MISSILE_1
: 9
.....SCANNING SENSOR TRACK ESTABLISHED
.....MISSILE_1 IDENTIFIED AS IR_MISSILE
.....TGO UPDATED ON MISSILE_1
.....COUNTERMEASURE APPLIED TO MISSILE_1
: 10
: 11
.....SCANNING SENSOR RETURNED TO SEARCH
      MODE
: 12
: 13
: 14
: 15
EMITTER 1 INACTIVE
.....ECM ON EMITTER 1 TURNED OFF AT
      TIME 15
: 16
EMITTER 2 INACTIVE
.....ECM ON EMITTER 2 TURNED OFF AT
      TIME 16
: 17
END OF SIMULATION
```


REFERENCES

[1]   Hayes-Roth, F, Donald A. Waterman
      and Douglas B. Lenat (eds.)
      Building ExpertSystem, Volume 1,
      First Edition, Addison Wesley,
      1983, PP. 183-186.

[2]   Forgy, Charles L., OPS5 User's
      Manual, Carnegie Mellon University,
      1981.

NOTES